INTERDISCIPLINARY COMPUTING IN JAVA PROGRAMMING

THE KLUWER INTERNATIONAL SERIES
IN ENGINEERING AND COMPUTER SCIENCE

INTERDISCIPLINARY COMPUTING IN JAVA PROGRAMMING

by

Sun-Chong Wang
TRIUMF, Canada

SPRINGER SCIENCE+BUSINESS MEDIA, LLC

Library of Congress Cataloging-in-Publication

INTERDISCIPLINARY COMPUTING IN JAVA PROGRAMMING
by Sun-Chong Wang
ISBN 978-1-4613-5046-0 ISBN 978-1-4615-0377-4 (eBook)
DOI 10.1007/978-1-4615-0377-4

Originally published by Kluwer Academic Publishers in 2003
Softcover reprint of the hardcover 1st edition 2003

Printed on acid-free paper.

Contents

Preface xi

Part I Java Language

1. JAVA BASICS 3
   - 1.1 Object Oriented Programming 3
   - 1.2 An Object Example 4
   - 1.3 Primitive Data Types 8
   - 1.4 Class Constructor 9
   - 1.5 Methods of a Class 9
   - 1.6 Exceptions 10
   - 1.7 Inheritance 10
   - 1.8 Usage of the `Matrix` Class 11
   - 1.9 Running the Program 13
   - 1.10 Summary 14
   - 1.11 References and Further Reading 15
2. GRAPHICAL AND INTERACTIVE JAVA 17
   - 2.1 Windowed Programming 17
   - 2.2 Example of a Window Object 18
   - 2.3 Frame 23
   - 2.4 Panel 23
   - 2.5 Menu 24
   - 2.6 Interactions 24
   - 2.7 File Input/Output 24

2.8 StreamTokenizer 26
2.9 Graphics 27
2.10 Printing 36
2.11 Summary 37
2.12 References and Further Reading 37

3. HIGH PERFORMANCE COMPUTING 39
3.1 Parallel Computing 39
3.2 Java Threads 40
3.3 An Example of Parallel Computing 41
3.4 Distributed Computing 41
3.5 Remote Method Invocation 41
3.6 An RMI Client 42
3.7 The Remote Interface 44
3.8 Serialization 48
3.9 A Reflective RMI Server 48
3.10 Reflection 51
3.11 Build and Run the Server 51
3.12 Build and Run the Client 53
3.13 Summary 54
3.14 Appendix 55
3.15 References and Further Reading 55

Part II Computing

4. SIMULATED ANNEALING 59
4.1 Introduction 59
4.2 Metropolis Algorithm 60
4.3 Ising Model 61
4.4 Cooling Schedule 62
4.5 3-Dimensional Plot and Animation 62
4.6 An Annealing Example 63
4.7 Minimization of Functions of Continuous Variables 77
4.8 Summary 78
4.9 References and Further Reading 79

5. ARTIFICIAL NEURAL NETWORK 81
5.1 Introduction 81
5.2 Structural vs. Temporal Pattern Recognition 84
5.3 Recurrent Neural Network 84
5.4 Steps in Designing a Forecasting Neural Network 86
5.5 How Many Hidden Neurons/Layers ? 87
5.6 Error Function 87
5.7 Kohonen Self-Organizing Map 88
5.8 Unsupervised Learning 89
5.9 A Clustering Example 90
5.10 Summary 99
5.11 References and Further Reading 100

6. GENETIC ALGORITHM 101
6.1 Evolution 101
6.2 Crossover 102
6.3 Mutation 103
6.4 Selection 104
6.5 Traveling Salesman Problem 105
6.6 Genetic Programming 113
6.7 Prospects 114
6.8 Summary 115
6.9 References and Further Reading 115

7. MONTE CARLO SIMULATION 117
7.1 Random Number Generators 117
7.2 Inverse Transform Method 118
7.3 Acceptance-Rejection (Von Neumann) Method 119
7.4 Error Estimation 120
7.5 Multivariate Distribution with a Specified Correlation Matrix 121
7.6 Stochastic-Volatility Jump-Diffusion Process 122
7.7 A Cash Flow Example 123
7.8 Variance Reduction Techniques 130
7.9 Summary 130
7.10 References and Further Reading 131

8. MOLECULAR DYNAMICS 133
8.1 Computer Experiment 133
8.2 Statistical Mechanics 134
8.3 Ergodicity 135
8.4 Lennard-Jones Potential 135
8.5 Velocity Verlet Algorithm 137
8.6 Correcting for Finite Size and Finite Time 138
8.7 An Evaporation Example 139
8.8 Summary 144
8.9 References and Further Reading 145

9. CELLULAR AUTOMATA 147
9.1 Complexity 147
9.2 Self-Organized Criticality 148
9.3 Simulation by Cellular Automata 148
9.4 Lattice Gas Automata 149
9.5 A Hydrodynamic Example 151
9.6 Summary 164
9.7 References and Further Reading 165

10. PATH INTEGRAL 167
10.1 Feynman's Sum Over Histories 167
10.2 Numerical Path Integration and Feynman-Kac Formula 170
10.3 Options in Finance 171
10.4 A Path Integral Approach to Option Pricing 171
10.5 Importance Sampling (Metropolis-Hastings algorithm) 172
10.6 Implementation 174
10.7 Summary 179
10.8 References and Further Reading 180

11. DATA FITTING 181
11.1 Chi-Square 181
11.2 Marquardt Recipe 182
11.3 Uncertainties in the Best-Fit Parameters 183
11.4 Arbitrary Distributions by Monte Carlo 183
11.5 A Surface Fit Example 187

11.6 Summary 193
11.7 References and Further Reading 194

12. BAYESIAN ANALYSIS 195
12.1 Bayes Theorem 195
12.2 Principle of Maximum Entropy 196
12.3 Likelihood Function 197
12.4 Image/Spectrum Restoration 198
12.5 An Iterative Procedure 201
12.6 A Pixon Example 202
12.7 Summary 209
12.8 References and Further Reading 210

13. GRAPHICAL MODEL 211
13.1 Directed Graphs 211
13.2 Bayesian Information Criterion 212
13.3 Kalman Filter 214
13.4 A Progressive Procedure 215
13.5 Kalman Smoother 217
13.6 Initialization of the Filter 218
13.7 Helix Tracking 218
13.8 Buffered I/O 221
13.9 The Kalman Code 226
13.10H Infinity Filter 235
13.11Properties of H Infinity Filters 236
13.12Summary 237
13.13References and Further Reading 238

14. JNI TECHNOLOGY 241
14.1 Java Native Interface 241
14.2 JNI HOW-TO 242
14.3 Call Fortran Programs from C 242
14.4 A JNI Example 244
14.5 Summary 251
14.6 References and Further Reading 251

Appendices 253

A 253

A.1 Web Computing 253

A.2 Class Sources 255

Index 261

Preface

This book is intended for personnel in quantitative research. Prerequisite is elementary calculus. Knowledge of probability and statistics is helpful but not mandatory. We introduce language and computing in separate parts of the book. Java is selected as the programming language because, among others, it is easy to learn. A computing methodology is elaborated in each chapter in part II. Each chapter contains a worked out example in Java.

In part I, the 1st chapter introduces Java elements, especially the concept of 'object'. We write a matrix object as an example. Chapter 2 is about graphics in Java and lays the foundation of the plotting and user-interaction programs for the rest of the book. Chapter 3 illustrates parallel and distributed computing in Java. We set up a client-server system as an example of distributed computing.

Chapters in part II are basically independent. Simulated annealing and genetic algorithm are usually applied in optimization problems. We find the ground state energy of a physical system as an example for simulated annealing. Traveling salesman problem is solved using genetic algorithm. Artificial neural network is used in AI and control. We apply neural network to perform clustering as the example. Monte Carlo method, molecular dynamics, and cellular automata are seen in simulation in different fields. We simulate stock prices using Monte Carlo and particle motions using molecular dynamics and cellular automata. Feynman's path integral is an alternative evaluation method in physics and chemistry. We price financial derivatives by path integral as an illustration. Chi-square fitting is indispensable in data analysis. We present a two dimensional surface fitting in the chapter example. Bayesian technique is gaining momentum as computing power advances. We deconvoluted a blurred image by Bayesian method. Graphical models are applied to decision making. We track particle trajectories in space using Kalman filter as a graphical model example. JNI technology allows Java programmers to call methods written in other languages. We demonstrate the way to call Fortran library routines from

Java. In the appendix, we show how to transform standalone Java applications into applets to realize web computing.

Source codes for an application in each chapter are typically organized in the following way: a class, holding a main method, sets up a frame, a pair of classes that are responsible for observing and drawing changes in data, a class creating a dialog box for user input and program output, and finally an 'engine' class that performs computing of the methodology of the chapter. In Chapter 4 (and appendix), we show all the source codes for the application. Starting from Chapter 5, we then show only the engine class to save space. A reader can easily create an application that have the plot and dialog box as shown in each chapter by simply revising the complete sources of Chapter 4.

I am indebted to Drs. Nathan Rodning of University of Alberta, Chary Rangacharyulu of University of Saskatchewan, Mamoru Fujiwara of Osaka University, Henry Tsz-king Wong of Academia Sinica, George Seidel of Brown University, Noel Coron of University of Paris XI, and Hinko Henry Stroke of New York University for enlightening discussions. I thank Susan Lagerstrom-Fife and Sharon Palleschi of Kluwer Academic Publishers for editorial assistance. This book is dedicated to Helen Weiwen Wang and my parents.

SUN-CHONG WANG

PART I

JAVA LANGUAGE

Chapter 1

JAVA BASICS

Java[1] programming language, espousing the tenet of *object orientation*, evolves in the time of the Internet and has been gaining much popularity among programmers. We start in this chapter introducing Java syntax by elaborating on an object example.

1.1 Object Oriented Programming

An object can be thought of as an individual in an organization. The state of the individual at a particular instance of time is characterized by her variables as well as constants. The individual performs her duties with her methods, which, among other things, can alter her state. When groups of individuals are orchestrated under a management, tasks are engaged and goals of the organization are reached. Likewise, in Java programming, we write a file, say, `Engineers.java`, in which variables and constants are defined and methods are implemented. The file is a *class*, as suggested by the suffix `.class` after it is successfully compiled. One or several *object*s of the same class can then be created by *instantiation* (We will detail what is meant by instantiation later in this chapter). In analogy, a class can be imagined as a position in an organization. Once a class (position) is devised, objects (personnel) can be instantiated (employed) to perform the tasks specified by the class. Finally, a master class, which contains the `main()` method, commands all the other classes (`Engineers.class`, `Sales.class`, ...) in the application, serving as top management in a hierarchical organization.

Object oriented programming was inspired by the way real world works, as experienced by everyone in mundane life. In traditional computer languages,

[1] Java is a trademark of Sun Microsystems Inc.

data, variables, and methods (subroutines) are separated. In object oriented languages, such as Java, relevant variables and methods are bundled into an object. An object, considered as a self-contained module, can be repeatedly used (and therefore tested). Re-usability of consistent objects greatly shortens the time for software design and development. Although programs in conventional languages like C and Fortran 90 can be dressed with a flavor of object-oriented paradigm, the syntax of Java makes transformation of the concept into coding straightforward. Other important features of Java, such as graphics, interactivity, parallel and distributed computing, are topics of following chapters.

1.2 An Object Example

Listing 1.1 shows the content of the file `Matrix.java`.

Java is case sensitive. Statements can start from any row or column in the program file. A statement terminates with a semicolon ;. A paired curly braces { and } bound the body of a collection of statements.

Comments in Java reside between /* and */. A short comment often comes after the double slash //.

In Listing 1.1, the keyword `public class` defines the class `Matrix`, which should also be the name of the file. `public` here means this `Matrix` class is callable by all other classes. The statement `import java.lang.*;` in the beginning indicates this file needs support from objects in the standard Java library package named `java.lang`. Those standard Java packages are installed when the Java Development Kit is installed in your system. We will need to import many other useful standard Java libraries as we proceed in later chapters.

```
/*      Sun-Chong Wang
        TRIUMF
        4004 Wesbrook Mall
        Vancouver, V6T 2A3
        Canada
        e-mail: wangsc@triumf.ca

        Matrix.java performs matrix algebra including
        inversion */

import java.lang.*;

public class Matrix {
 double[][] M; // array to store the matrix elements
 double det;    // determinant

 public Matrix(int row, int col) {
        M = new double[row][col];
        det = 0.0;
 }      // end of constructor

 public double[][] plus(double[][] N) {
 // M[][] of this object + input N[][]
        if ((M.length == N.length) && (M[0].length == N[0].length)) {
           double[][] tmp = new double[M.length][M[0].length];
           for (int i=0; i<M.length; i++) {
```

```
                for (int j=0; j<M[0].length; j++) {
                    tmp[i][j] = M[i][j] + N[i][j];
                }
            }
            return tmp;
        } else {
            System.out.println("Error in matrix addition");
            return null;
        }
}

public double[][] minus(double[][] N) {
// M[][] of this object - input N[][]
        if ((M.length == N.length) && (M[0].length == N[0].length)) {
            double[][] tmp = new double[M.length][M[0].length];
            for (int i=0; i<M.length; i++) {
                for (int j=0; j<M[0].length; j++) {
                    tmp[i][j] = M[i][j] - N[i][j];
                }
            }
            return tmp;
        } else {
            System.out.println("Error in matrix subtraction");
            return null;
        }
}

public double[][] times(double[][] N) {
// M[][] of this object x input N[][]
        if (M[0].length == N.length) {
            double[][] tmp = new double[M.length][N[0].length];
            for (int i=0; i<M.length; i++) {
                for (int j=0; j<N[0].length; j++) {
                    tmp[i][j] = 0.0;
                    for (int k=0; k<M[0].length; k++) {
                        tmp[i][j] += M[i][k] * N[k][j];
                    }
                }
            }
            return tmp;
        } else {
            System.out.println("Error in matrix multiplication");
            return null;
        }
}

public double[][] transpose() {
// transpose M[][]
        double[][] tmp = new double[M[0].length][M.length];
        for (int i=0; i<M[0].length; i++) {
            for (int j=0; j<M.length; j++) {
                tmp[i][j] = M[j][i];
            }
        }
        return tmp;
}

public double[][] ret_inv() throws MyMatrixExceptions {
// returns the inverse of M[][]
        double[][] save = new double[M.length][M[0].length];
        double[][] tmp;

        for (int i=0; i<M.length; i++)
            for (int j=0; j<M[0].length; j++)
```

```
                save[i][j] = M[i][j];

        inverse();
        tmp = M;
        M = save;
        return tmp;
}

public void inverse() throws MyMatrixExceptions {
// inplace matrix inversion with full pivoting
// also calculates the determinant
        int i, j, k;
        int iext=0, jext=0, itemp, jtemp;
        int nmax = M.length;
        int[] ir = new int[nmax];
        int[] ic = new int[nmax];
        double aext, atemp, de;

        de=1.0;
        for (j=0;j<nmax;j++) {
            ic[j] = j;
            ir[j] = j;
        }
        for (k=0;k<nmax;k++) {
            aext=0.0;
            for (i=k;i<nmax;i++)
                for (j=k;j<nmax;j++)
                    if (aext<Math.abs(M[i][j])) {
                       iext=i;
                       jext=j;
                       aext=Math.abs(M[i][j]);
                    }
            if (aext<=0.0)
            throw new MyMatrixExceptions("Error in matrix inversion 1");
            if (k!=iext) {
               de = -de;
               for (j=0;j<nmax;j++) {
                   atemp=M[k][j];
                   M[k][j]=M[iext][j];
                   M[iext][j]=atemp;
               }
               itemp=ic[k];
               ic[k]=ic[iext];
               ic[iext]=itemp;
            }
            if (k!=jext) {
               de = -de;
               for (i=0;i<nmax;i++) {
                   atemp=M[i][k];
                   M[i][k]=M[i][jext];
                   M[i][jext]=atemp;
               }
               itemp=ir[k];
               ir[k]=ir[jext];
               ir[jext]=itemp;
            }
            aext=M[k][k];
            de*=aext;
            M[k][k]=1.0;
            for (j=0;j<nmax;j++) M[k][j]/=aext;
            for (i=0;i<nmax;i++)
                if (k!=i) {
                   aext=M[i][k];
                   if (aext!=0.0) {
                      M[i][k]=0.0;
                      for (j=0;j<nmax;j++) M[i][j]-=aext*M[k][j];
```

```
                }
            }
        }       // k loop
        int idim = nmax-1;
        for (k=0;k<idim;k++) {
            int kk=k+1;
            if (k!=ic[k]) {
               for (i=kk;i<nmax;i++) if (k==ic[i]) break;
               if (i == nmax) throw new
                  MyMatrixExceptions("Error in matrix inversion 2");
               for (j=0;j<nmax;j++) {
                   atemp=M[j][k];
                   M[j][k]=M[j][i];
                   M[j][i]=atemp;
               }
               itemp=ic[i];
               ic[i]=ic[k];
               ic[k]=itemp;
            }
            if (k!=ir[k]) {
               for (j=kk;j<nmax;j++) if (k==ir[j]) break;
               if (j == nmax) throw new
                  MyMatrixExceptions("Error in matrix inversion 3");
               for (i=0;i<nmax;i++) {
                   atemp=M[k][i];
                   M[k][i]=M[j][i];
                   M[j][i]=atemp;
               }
               itemp=ir[j];
               ir[j]=ir[k];
               ir[k]=itemp;
            }
        }   // k loop
        det=de;
}

public double[][] rotation(int k, double angle) {
// rotate an angle along k axis
        int i,j;

        if (k<3) {
           i = (k+1)%3;
           j = (k+2)%3;
           double[][] tmp = new double[3][3];

           for (int row=0; row<3; row++) {
               for (int col=0; col<3; col++) {
                   if (row != col) {
                      tmp[row][col] = 0.0;
                   } else {
                      tmp[row][col] = 1.0;
                   }
               }
           }
           tmp[i][i] = Math.cos(angle);
           tmp[j][j] = Math.cos(angle);
           tmp[i][j] = Math.sin(angle);
           tmp[j][i] = -tmp[i][j];

           return tmp;
        } else {
           System.out.println("Error in rotational matrix");
           return null;
        }
}
}       // end of class Matrix
```

Listing 1.1 Matrix.java

1.3 Primitive Data Types

Table 1.1. Primitive data types in Java. Their size/format, minimum, and maximum values accessible via `Type.MIN_VALUE` and `Type.MAX_VALUE`, where `Type` can be `Byte`, `Short`, `Integer`, `Long`, `Float`, or `Double`.

primitive type	*size*	MIN_VALUE	MAX_VALUE
byte	8-bit	-128	127
short	16-bit	-32768	32767
int	32-bit	-2147483648	2147483647
long	64-bit	-9223372036854775808	9223372036854775807
float	32-bit IEEE 754	1.4E-45	3.4028235E38
double	64-bit IEEE 754	4.9E-324	1.7976931348623157E308
char	16-bit Unicode		
boolean	true or false		

Within the `Matrix` class, first of all, a 2-dimensional double array is declared. Primitive data types in Java include `boolean`, `char`, `byte`, `short`, `int`, `long`, `float`, and `double`. Their representations and ranges are shown in Table 1.1. To create an instance of integer `myInt`, the following statement is used,

```
int myInt;
```

An array, unlike primitive data types, assumes the status of a class. The statement in `Matrix.java`,

```
double[][] M;
```

therefore declares that `M` is a class of a 2-dimensional array whose elements are of type `double`. Before `M` can be deployed, it has to be instantiated; namely, enough memory space has to be allocated to store the content of the array. This act of instantiation is accomplished by the statement with the keyword `new` like this,

```
M = new double[3][3];
```

where an instance of a 3 by 3 `double` array is created. Variables defined in this *field* of a class are accessible by all the methods defined in the class. A *method* in fact implements the way the state (variables) of the object is changed upon requests by foreign objects.

1.4 Class Constructor

In the spirit of array creation by calling `new double[3][3]`, we have to write a constructor for the class `Matrix`. The constructor, bearing the same name as the class name, is usually the first method of the class, as in the `Matrix` example of Listing 1.1. In this simple example, two integers, `row` and `col` in the argument list of the constructor method, are passed and used to specify the dimensional lengths of the array. To create an instance of the class `Matrix` in other files, a `new` command is issued after `Matrix` declaration:

```
Matrix A, B;
A = new Matrix(4,4);
B = new Matrix(3,4);
```

The variable array `M`'s in `A` and `B` can then be accessed via,[2]

```
A.M[0][1] = 2.0;
B.M[2][3] = -4.0;
```

and so on. Note array indexes start from 0 in Java as in C/C++.

We have so far done nothing more than creating arrays. The responsibilities of an object and thus how it interacts with other objects are defined in its methods.

1.5 Methods of a Class

Following the constructor method in the class `Matrix` is implemented the method `plus()`. The declaration of the method,

```
public double[][] plus(double[][] N)
```

says that this method is public, thereby can be called by any other objects. `double[][]` after the `public` keyword depicts that this method returns a 2-dimensional array when called. The `double[][] N` inside the parentheses indicates that this method accepts a 2-dimensional array as input. It is to be noted that, in Java, arguments of primitive types are passed by *value* and that arguments of types other than primitive are passed by *reference* (pointer or memory address). That is, any arguments created by the `new` keyword are passed by reference.

Inside the body of the method, delimited by curly braces { and }, is implemented how the input 2-dimensional array `N` is added to the 2-dimensional array, `M`, of the current matrix object:

$$\mathtt{tmp}_{ij} = \mathtt{M}_{ij} + \mathtt{N}_{ij}$$

[2]Note that, to preserve encapsulation of data in an object, an object-oriented purist may prefer methods like `A.setValueAt(0,1,2.0)` and `B.setValueAt(2,3,-4.0)` to alter variables of the objects. Method-callings, however, take a longer time than statements. In some cases, we simply optimize speed at the cost of object encapsulation.

The range of i and j is between 0 and `M.length-1` inclusive. `M.length` returns the length of the first dimension of the array. Likewise, the length of the second dimension of the array is kept in the constant `M[0].length`.

`int i,j`, and `double[][] tmp` defined in this method are temporarily created by the compiler for the interim task within the block of this method. They will be (hopefully) garbage collected by the compiler when they are no longer in use (such as after the method is exited).

It is perhaps most convenient for many programmers that the looping and logical syntax of Java look much similar to those of C and C++. Java also adopts some of the shorthands of C. For example, the statement `value++;` directs the program to deliver the value of `value` and then to increment `value` by 1. `++value;` increments `value` by 1 before delivering the value of `value`. `value /= 10;` is a shortcut to `value = value/10;`. In fact, most numerical routines in C and C++ can be ported to Java without much change. In addition, Java relieves C/C++ programmers from array memory allocation and deallocation chores. The array length attribute in Java also facilitates array bound checks.

Subsequent methods `minus()`, `times()`, `transpose()` in the class `Matrix` are no more particular than `plus()`. We will demonstrate their usage later in the chapter. We mention briefly here that `System.out.println("string");` writes `string` to the standard output of the computer, usually the monitor screen of the computer. More I/O utilities in Java will be introduced in the next chapter.

1.6 Exceptions

The `inverse()` method takes no input parameters. It inverts the 2 dimensional array variable `M` of the `Matrix` object. The algorithm used to invert the matrix is the familiar Gauss-Jordan elimination (with full pivoting) method found in most texts on numerical computation.

We now encounter in this method a handy utility of Java called *exceptions*. It may happen that some matrices cannot be inverted. For example, the set of linear equations corresponding to the matrix equation does not have a solution. In this case, the `inverse()` method will fail and it is desirable that the failure be handled gracefully without aborting program execution. To accomplish this, the method indicates that it throws `MyMatrixExceptions`, which is a class inheriting Java's class `Throwable`. Inheritance is another feature of object oriented Java and will be addressed in the next section. Examining the algorithm, we observe that unsuccessful inversions occur when for example numbers are divided by zero. Instances of the class `MyMatrixExceptions` are created and thrown in these occasions. The exceptions are then caught in the `try-catch` block in the object that calls the `inverse()` method. Examples of the try catch block will be seen shortly in the following section.

The difference between methods `ret_int()` and `inverse()` is in their return types.

1.7 Inheritance

A look at Listing 1.2 shows how easily `MyMatrixExceptions` inherits Java's `Throwable` class by simply `extends Throwable`. Again, the first method, bearing the same name as the class (`MyMatrixExceptions`), is the constructor of the class. It accepts a single `String` object as the input parameter. Here super means the parent of MyMatrixExceptions which is Throwable.

```
/*      Sun-Chong Wang
        TRIUMF
        4004 Wesbrook Mall
        Vancouver, V6T 2A3
        Canada
        e-mail: wangsc@triumf.ca

        MyMatrixExceptions.java reports failure to
        the calling object */

class MyMatrixExceptions extends Throwable {
 public MyMatrixExceptions(String s) {
        super(s);
 }      // end of constructor
}       // end of class
```

Listing 1.2 MyMatrixExceptions.java

1.8 Usage of the Matrix Class

The remaining method rotation() in Matrix.java performs coordinate system transformations along axes. It will not be used until we rotate a 3-dimensional geometrical object in Chapter 4.

File MatrixDemo.java in Listing 1.3 demonstrates how Matrix objects are used. Since MatrixDemo is itself a class, an instance of it has to be created before the variables and methods contained in it can be maneuvered. Instantiation of MatrixDemo is done in the main() method.

```
public static void main(String args[])
```

static here means there can only be one such method in possibly many instances of MatrixDemo. This makes sense since one entry point to the application is necessary and sufficient. void indicates that the method returns nothing. args[] is a String array storing command line arguments similar to that in C.

```
/*      Sun-Chong Wang
        TRIUMF
        4004 Wesbrook Mall
        Vancouver, V6T 2A3
        Canada
        e-mail: wangsc@triumf.ca

        MatrixDemo.java demonstrates use of the Matrix class */

import java.lang.*;

class MatrixDemo {
 final int Size = 3;
 Matrix A = new Matrix(Size,Size);
 Matrix B = new Matrix(Size,Size);
 Matrix C = new Matrix(Size,Size);
 Matrix D = new Matrix(Size,Size);
 Matrix E = new Matrix(Size,Size);

 public static void main(String args[]) {
        MatrixDemo demo = new MatrixDemo();
```

```
        System.out.println("A = ");
        demo.printMatrix(demo.A);

        System.out.println("B = ");
        demo.printMatrix(demo.B);

        demo.C.M = demo.A.transpose();
        System.out.println("transpose of A = ");
        demo.printMatrix(demo.C);

        demo.C.M = demo.A.plus(demo.B.M);
        System.out.println("A + B = ");
        demo.printMatrix(demo.C);

        demo.C.M = demo.B.minus(demo.A.M);
        System.out.println("B - A = ");
        demo.printMatrix(demo.C);

        try {
             demo.C.M = demo.A.ret_inv();
             System.out.println("inverse of A = ");
             demo.printMatrix(demo.C);
        } catch (MyMatrixExceptions mme) {
             System.out.println(mme.getMessage());
        }

        demo.D.M = demo.A.times(demo.C.M);
        System.out.println("A x A_inverse = ");
        demo.printMatrix(demo.D);

        try {
          demo.E.M =
          demo.B.minus(demo.A.times(demo.C.plus(demo.A.ret_inv())));
          System.out.println("B - (A x (A_inverse + A_inverse)) = ");
          demo.printMatrix(demo.E);
        } catch (MyMatrixExceptions mme) {
          System.out.println(mme.getMessage());
        }
 }      // end of main

 public MatrixDemo() {
        A.M[0][0] = 1.; A.M[0][1] = 2.; A.M[0][2] = 5.;
        A.M[1][0] = 2.; A.M[1][1] = 3.; A.M[1][2] = 8.;
        A.M[2][0] = -1.;A.M[2][1] = 1.; A.M[2][2] = 2.;

        B.M[0][0] = 1.; B.M[0][1] = 0.; B.M[0][2] = 0.;
        B.M[1][0] = 0.; B.M[1][1] = 1.; B.M[1][2] = 0.;
        B.M[2][0] = 0.; B.M[2][1] = 0.; B.M[2][2] = 1.;
 }      // class constructor

 public void printMatrix(Matrix C) {
        for (int i=0; i<Size; i++) {
            for (int j=0; j<Size; j++)
                 System.out.print(C.M[i][j]+" ");
            System.out.println("");
        }
        System.out.println("");
 }
}       // end of class MatrixDemo
```

Listing 1.3 MatrixDemo.java

In the variable field of the class, five `Matrix` variables `A, B, C, D, E` are declared and instantiated to be 3 by 3 in single statements. The keyword `final` in,

```
final int Size = 3;
```

modifies the property of the `int` so that `Size` is now a constant integer with a fixed value of 3.

The `main()` method is followed by the class constructor, `MatrixDemo()`, which implicitly calls the default (parent) constructor. Note that every class in Java has its immediate superclass. At the top of the class hierarchy is a class called `Object`. The constructor `super()` is the constructor of the ancestral class.

Next in the class constructor, values of the array elements are assigned. All the `MatrixDemo` methods are called within the `main()` after an instance of `MatrixDemo` is realized,

```
MatrixDemo demo = new MatrixDemo();
```

Recall that the keyword `new` calls the constructor of the class. `Matrix` objects `A`, `B`, `C`, `D`, and `E` are now made to interact by performing subtraction, addition, multiplication between them and transpose and inversion on itself. The results are printed out on the screen by the method `printMatrix()` of `demo`, for example,

```
demo.printMatrix(demo.C);
```

Note that variables and methods of an object are referenced via the `.` operator, as in the above example. The `try-catch` block encompassing the matrix inversion method is also noted. This block is mandatory since the method `inverse()` indicates that it may throw exceptions when occasions arise. Compilation will fail if the `try-catch` block is missing. When an exception does happen, it's caught by `catch` and the warning message can be printed out. Namely, remedial procedures are taken in the `catch` and execution proceeds to the next statement without crashing program running.

1.9 Running the Program

In the present working directory we now have three files: `Matrix.java`, `MatrixDemo.java`, and `MyMatrixExceptions.java`. Before compiling, we set up the environment under the system prompt $ by,

```
$export JAVA_HOME=/home/wangsc/JAVA/jdk1.2.2
$export PATH=$JAVA_HOME/bin:$PATH
$export CLASSPATH=.
```

The reader should replace the above Java home directory with the directory containing the Java tools in her system. Setting-up of the environment is done only once (per login). We can then compile the sources by the Java compiler, `javac`,

```
$javac MatrixDemo.java
```

We observe that three (bytecode) class files have been created by the compiler. We now launch the application by the Java interpreter (or launcher), `java`,

```
$java MatrixDemo
```

We immediately get the following output on screen,

```
A =
1.0 2.0 5.0
2.0 3.0 8.0
-1.0 1.0 2.0

B =
1.0 0.0 0.0
0.0 1.0 0.0
0.0 0.0 1.0

transpose of A =
1.0 2.0 -1.0
2.0 3.0 1.0
5.0 8.0 2.0

A + B =
2.0 2.0 5.0
2.0 4.0 8.0
-1.0 1.0 3.0

B - A =
0.0 -2.0 -5.0
-2.0 -2.0 -8.0
1.0 -1.0 -1.0

inverse of A =
1.9999999999999996 -0.9999999999999999 -0.9999999999999999
11.999999999999998 -7.0 -1.9999999999999998
-4.999999999999999 3.0000000000000004 0.9999999999999999

A x A_inverse =
1.0 1.7763568394002505E-15 0.0
0.0 1.0000000000000036 0.0
0.0 8.881784197001252E-16 0.9999999999999999

B - (A x (A_inverse + A_inverse)) =
-1.0 -3.552713678800501E-15 0.0
0.0 -1.000000000000007 0.0
0.0 -1.7763568394002505E-15 -0.9999999999999998
```

The matrix class of this chapter exemplifies creation and use of objects in Java. In part II, we will meet occasions where we need this matrix class. Java also provides a mathematics class, `java.lang.Math`, that performs square root, calculates sines, cosines, and so on. Visit Sun Microsystems' website for online documentations of all the classes in a Java distribution: `java.sun.com`.

1.10 Summary

As software becomes more complicated, expenses on the maintenance sky-rocket. Besides the urgency to develop intelligent and autonomic software that

MatrixDemo.java ——— Matrix.java ——— MyMatrixExceptions.java

Figure 1.1. Source programs in the matrix object demonstration

can maintain and heal itself and each other, a cross-platform programming language is an advantage. Java was introduced with such an idea of 'write once, run everywhere'.

We introduced the concept of object oriented programming. A class is a blueprint that specifies the functionality. Once a blueprint is laid out, instances of the class can be incarnated via the `new` statement, which in fact calls the constructor method of the class.

An `int (long)` and `float (double)` in Java are represented by 4 (8) bytes. Arrays in Java are objects and their instantiation and initialization are by the `new` statements. Information on the array length is retrievable with the array name.

All objects in Java, including the ones the programmer writes, are subclasses. The inheritance property makes it easy for a programmer to use classes written by others. For example, one may find the class `Matrix` in this chapter useful but wants to add to it her own methods. She can then simply `extends Matrix` and work on her supplements.

1.11 References and Further Reading

M. Smith, "Java: an Object-Oriented Language", McGraw-Hill International (UK) Limited, London (1999)

G. Cornell and C.S. Horstmann, "Core Java", Prentice Hall, NJ (1996)

C. Laffra, "Advanced Java", Prentice Hall, NJ (1997)

D. Flanagan, "Java in a Nutshell: A Desktop Quick Reference", third ed. O'Reilly & Associates (1999)

Chapter 2

GRAPHICAL AND INTERACTIVE JAVA

After numerical calculations, it is often desirable that relations of the numbers are displayed in charts, curves, histograms, contour plots, or any other graphical forms. In traditional languages like Fortran or C, we usually have to resort to some plotting tools available in the host system. In this scenario, not only do we have to learn the the graphics package, but also are concerned about the input/output formats between various tools. In contrast, Java comes with a rich supply of graphics classes. Results of numerical objects can be passed around and readily plotted by the graphics objects within the same application. Seamless integration of the two operations is another advantage of Java.

Furthermore, Java provides a handful of basic geometrical objects that ease programmers' graphing travails. In this chapter, we show graphical user interaction using Java.

2.1 Windowed Programming

Figure 2.1 is the screen shot of the window we are going to create. It is the central topic of this chapter. Interactions of the program with the user are through the mouse, which is a common equipment of any computer besides the keyboard. When we click on one of the items on the menu bar, for example, `file`, a pull down submenu which contains more selective options will appear. Under the `file` menu, we see `Open` file, `Save` file, and `Quit` button in a column as shown in Figure 2.2. When the `Open` button is selected, a dialog window will pop up, prompting the user to select a file, in the current directory, for reading. The above scenario of interactivity is commonplace in modern software. Java has more than enough such classes for windowed programming. All we need to do is simply to inherit those window classes.

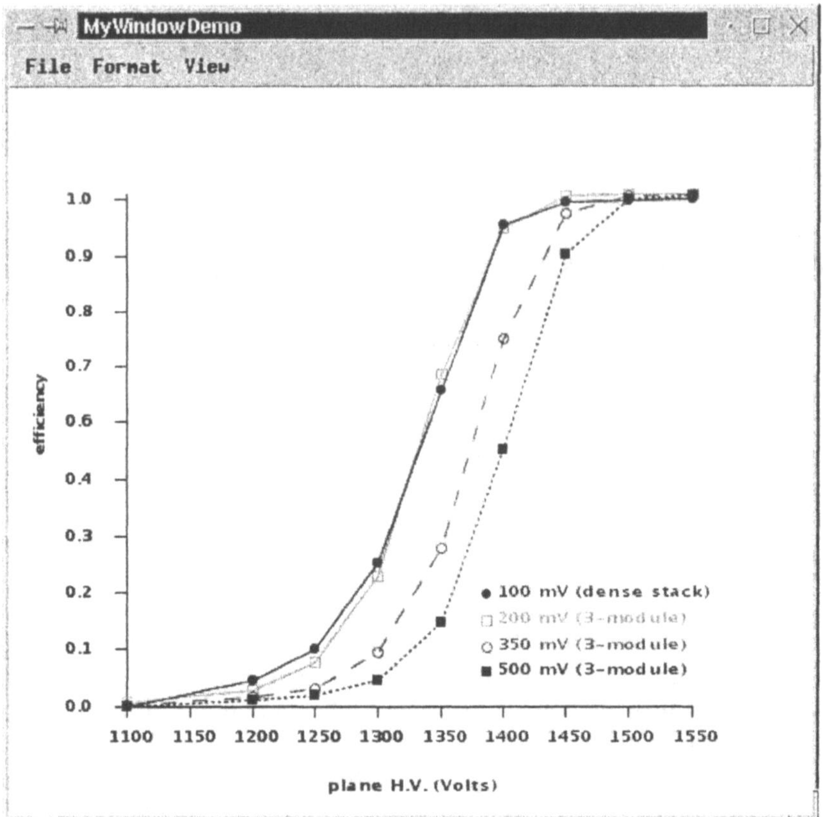

Figure 2.1. The window object

2.2 Example of a Window Object

The code of the window class of Figure 2.1 is shown in Listing 2.1. We need window utilities from Java, therefore classes in the package `java.awt` are being imported.

```
/*      Sun-Chong Wang
        TRIUMF
        4004 Wesbrook Mall
        Vancouver, V6T 2A3
        Canada
        e-mail: wangsc@triumf.ca

        MyWindow.java demonstrates Java window programming */

import java.awt.*;
import java.io.*;
import java.lang.*;
import java.awt.event.*;
import java.awt.print.*;
import java.awt.print.PrinterJob;
import java.text.DecimalFormat;
```

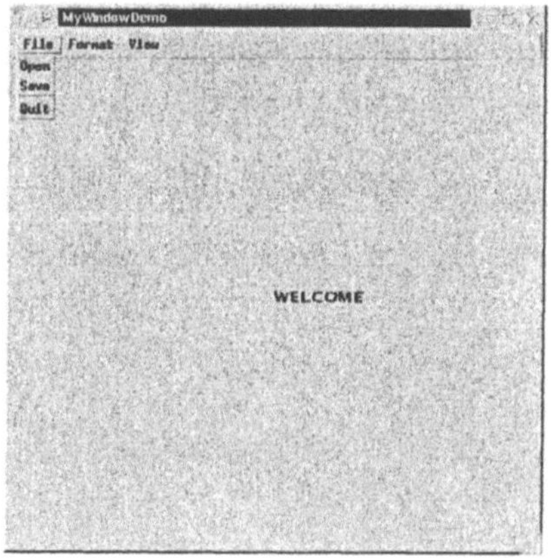

Figure 2.2. Items on the `File` pull-down menu

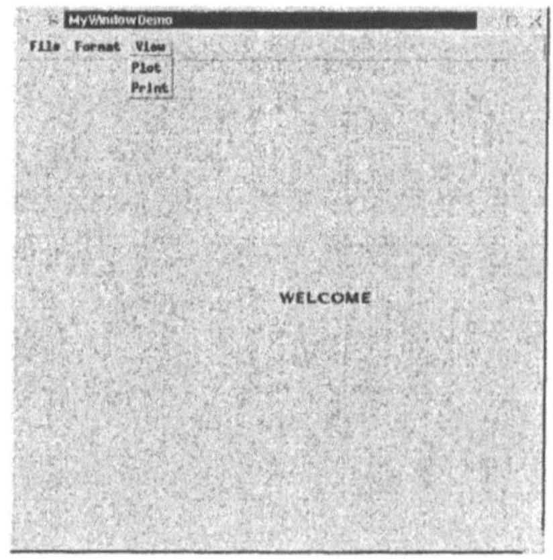

Figure 2.3. Items on the `View` pull-down menu

```
public class MyWindow extends Frame
                      implements ActionListener, Printable {
 Plotter plotting;
 int nColns, nLines, xIndex, nSkips;
 double[][] Table;
 boolean beforePlot = true;

 public static void main(String args[]) {
        MyWindow demo = new MyWindow();
        demo.show();
 }      // end of main

 public MyWindow() {
        super();
        setTitle("MyWindowDemo");

        nColns = 5;  // default input file format, no. of columns
        xIndex = 1;  // x column
        nSkips = 4;  // no. of lines to skip

        Font font = new Font("Dialog", Font.BOLD, 13);
        setFont(font);

        plotting  = new Plotter(this); // 'this' is MyWindow

        // enables quitting the application by clicking the
        // x button on the upper right corner of the window
        addWindowListener(new WindowAdapter() {
           public void windowClosing(WindowEvent e)
                  {System.exit(0);} });

        addMenus();

        Panel mypanel = new Panel();
        mypanel.setLayout(new BorderLayout());
        mypanel.add(plotting, BorderLayout.CENTER);
        add(mypanel);

        pack();
```

```
        setSize(new Dimension(500,500));
}       // end of constructor

private void addMenus() {
        MenuBar mymenubar = new MenuBar(); // list menus

        Menu myfile = new Menu("File");
        myfile.add("Open");    // items on this menu
        myfile.add("Save");
        myfile.addSeparator();
        myfile.add("Quit");

        Menu format = new Menu("Format");
        format.add("Import");

        Menu operate = new Menu("View");
        operate.add("Plot");
        operate.add("Print");

        myfile.addActionListener(this);
        format.addActionListener(this);
        operate.addActionListener(this);

        mymenubar.add(myfile);
        mymenubar.add(format);
        mymenubar.add(operate);
        setMenuBar(mymenubar);
}       // end of addMenus

// action handler
public void actionPerformed(ActionEvent e) {
        String action_is = e.getActionCommand();

        // action when a particular item is selected
        if (action_is.equals("Quit")) {
           System.exit(0);
        } else if (action_is.equals("Open")) {
           FileDialog opendlg = new FileDialog(this,
                                   "Open File",FileDialog.LOAD);
           opendlg.show();
           String infile = opendlg.getDirectory() + opendlg.getFile();
           if (opendlg.getFile() != null) {
              Message readingBox = new Message(this,"MyWindow",
                                          "Reading file...");
              readingBox.show();
              loadData(infile);
              readingBox.dispose();
           }
        } else if (action_is.equals("Import")) {
           FDialog formatdlg = new FDialog(this,"Format Dialog");
           formatdlg.show();
        } else if (action_is.equals("Save")) {
           FileDialog savedlg = new FileDialog(this,
                                "Save File As...",FileDialog.SAVE);
           savedlg.show();
           String outfile = savedlg.getDirectory()+savedlg.getFile();
           Message savingBox = new Message(this,"MyWindow",
                                        "Saving file...");
           savingBox.show();
           writeData(outfile);
           savingBox.dispose();
        } else if (action_is.equals("Plot")) {
           assignArrays();
```

```
            Message plottingBox = new Message(this,
                                     "MyWindow","Plotting file...");
            plottingBox.show();
            beforePlot = false;
            plotting.repaint();        // invokes paint()
            plottingBox.dispose();
        } else if (action_is.equals("Print")) {
            PrinterJob printJob = PrinterJob.getPrinterJob();
            PageFormat pf = new PageFormat();
            pf.setOrientation(pf.PORTRAIT);
            printJob.setPrintable(this,pf);
            if (printJob.printDialog()) {
               try {
                    printJob.print();
               } catch (Exception ex) {
                    ex.printStackTrace();
               }
            }
        }
}       // end of actionPorformed

public int print(Graphics g, PageFormat pf, int pi) {
// this method is required by the interface Printable
        if (pi >= 1) return Printable.NO_SUCH_PAGE;
        Graphics2D g2 = (Graphics2D) g;
        g2.translate(pf.getImageableX(), pf.getImageableY());
        plotting.paint(g2);
        return Printable.PAGE_EXISTS;
}

public void assignArrays() {
        int j = 0;

        plotting.x = Table[xIndex-1];
        plotting.y = new double[nColns-1][];

        for (int i=0; i<nColns; i++) {
            if (i != (xIndex-1)) {
               plotting.y[j] = Table[i];
               j += 1;
            }
        }
}

public void loadData(String infile) {
        int itmp;
        String stmp;

        nLines = 0;
        try {
             // file input, infile is from the dialog box
             FileInputStream fis  = new FileInputStream(infile);
             InputStreamReader br = new InputStreamReader(fis);
             LineNumberReader re = new LineNumberReader(br);

             itmp = -1;
             while (itmp != nLines) {
                    itmp = nLines;
                    stmp = re.readLine();
                    nLines = re.getLineNumber();
             }     // get the number of lines in the input file
             fis.close();
        } catch (IOException e) {
             System.out.println("IOExcpetion: "+e.getMessage());
```

```
        }

        Table = new double[nColns][nLines-nSkips];

        try {
            FileInputStream fis   = new FileInputStream(infile);
            InputStreamReader br = new InputStreamReader(fis);
            BufferedReader re    = new BufferedReader(br);
            StreamTokenizer sto  = new StreamTokenizer(re);

            // skip the first nSkips lines
            for (int i=0; i<nSkips; i++) stmp = re.readLine();

            for (int i=0; i<(nLines-nSkips); i++) {
                for (int j=0; j<nColns; j++) {
                    Table[j][i] = readNumber(sto);
                }
            }
            fis.close();
        } catch (FileNotFoundException fnfe) {
            System.out.println(fnfe.getMessage());
        } catch (IOException e) {
            System.out.println("IOExcpetion: "+e.getMessage());
        }
}       // end of readFile

// read numbers in plain or scientific notation
public double readNumber(StreamTokenizer sto) throws IOException {
        double output;
        Integer integer;

        sto.nextToken();
        output = sto.nval;
        sto.nextToken();
        if (sto.ttype == StreamTokenizer.TT_WORD) {
           if (sto.sval.length() > 1 &&
              sto.sval.substring(0,1).equalsIgnoreCase("E")) {
              integer = new Integer(sto.sval.substring(2));
              if (sto.sval.substring(1,2).equals("-")) {
                 output /= Math.pow(10.0, integer.doubleValue());
              } else {
                 output *= Math.pow(10.0, integer.doubleValue());
              }
           } else if (sto.sval.length() == 1 &&
              sto.sval.substring(0,1).equalsIgnoreCase("E")) {
                 sto.nextToken();      // get the + sign
                 sto.nextToken();      // get the exponent
                 output *= Math.pow(10.0, sto.nval);
           } else System.out.println("Error in the number format");
        } else
           sto.pushBack();

        return output;
}       // end of readNumber

public void writeData(String outfile) {
        try {
            // file output, outfile is from the dialog box
            FileOutputStream ostream = new FileOutputStream(outfile);
            PrintWriter      pw      = new PrintWriter(ostream);

            // set the decimal format of the numerals
            DecimalFormat df1 = new DecimalFormat("0000");
            DecimalFormat df2 = new DecimalFormat("0.0000");
```

```
            String fm = " ";

            for (int i=0; i<Table[0].length; i++) {
                for (int k=0; k<Table.length; k++) {
                    if (k == 0) fm = df1.format(Table[k][i]);
                    else fm = df2.format(Table[k][i]);
                    pw.print(fm+" ");
                }
                pw.println(" ");
            }

            pw.flush();
            ostream.close();
        } catch (IOException ee) {
            System.out.println("IOException : "+ee.getMessage());
        }
  }
}       // end of MyWindow class
```

Listing 2.1 MyWindow.java

2.3 Frame

This `MyWindow` class inherits, i.e. `extends`, the class `java.awt.Frame`. Here we have ignored `java.awt` since the `import` statement above told the compiler where to search for `Frame`. `ActionListener` is a category of classes called `interface`. In contrast to other object oriented programming languages like C++, multiple inheritance (inheritance from multiple parents) is not allowed in Java. One way to work around is via `implements`'ing interfaces. We therefore come up with the class declaration like this,

```
public class MyWindow extends Frame implements ActionListener {
```

We will encounter instances of `interface` in more detail in the next chapter. It suffices to say here that an `interface` is an abstract object where methods are only declared but not implemented. By `implements`'ing an interface, the programmer is obliged to provide her implementations of the methods.

Next in `MyWindow` comes the `main()` method, where an object of the class `MyWindow` is instantiated.

In the constructor of `MyWindow`, a font object is declared and instantiated. The method `setFont(font)` is not seen among the methods written below in `MyWindow`, indicating that it is a method of `Frame`, which `MyWindow` inherits. Through inheritance, we are exploiting Java utilities !

Next, more objects which are declared in the variable field are instantiated here in the constructor. `this` refers to self, i.e., `MyWindow`.

`pack()` is a method of `Frame`. `setSize(new Dimension(500,500))` defines the dimension of `MyWindow` in pixels.

`addMyMenu()` is one of the methods of this class, and is the subject of Section 2.5.

2.4 Panel

A `Panel` class provides space for any window component, including other panels. Here a panel object is declared and instantiated: `Panel mypanel = new`

`Panel();`. This panel object then invokes its `setLayout()` method to request an instance of `BorderLayout` to be the layout manager: `mypanel.setLayout (new BorderLayout());`. This layout manager, managing the space which is to be used by the `plotting` object, is then added to this panel: `mypanel.add (plotting, BorderLayout.CENTER);`. Finally this panel is added to the `MyWindow` object by `add(mypanel);`.

2.5 Menu

We might prefer separate menus for very different cuisines. We therefore in the method `addMyMenu()` first create an instance of `MenuBar` to hold subsequent submenus.

The `file` menu is then created: `Menu file = new Menu("File");`. An entrée in this menu is added: `file.add("Open")`. In this case, three entrées are available. This `file` menu is then added to the `menubar`. Other menus are populated in the same fashion. In the end of the method, the `menubar` is added to `MyWindow` by `setMenuBar(menubar)`. So far, when a menu is clicked by a customer, entrées are displayed. We then need a waiter to assist in receiving, placing, and delivering the order. The responsibilities of the waiter are prescribed by the interface `ActionListener` and implemented in the method `public void actionPerformed(ActionEvent e)`.

2.6 Interactions

An instance of `ActionEvent e` is passed as the argument for the method. Invoking the method `getActionCommand()` of `e` returns a `String` object holding the menu item clicked by the user. This item is to be identified in the `if else if else` block within which an appropriate action is performed in response to user selection.

We remind that these responding actions are included in the menus by the menu's `addActionListener()` method as in the statement `myfile.addActionListener(this)`. Again, `this` refers to `MyWindow`, which is capable of reacting because it implements `ActionListener`. Programmer's implementation of the method `actionPerformed()` in class `MyWindow` is mandatory since `MyWindow implements` the `ActionListener` interface.

2.7 File Input/Output

We now turn to one of the main subjects of any language, namely, input and output. Traditionally, we would issue a command from a UNIX or DOS shell like this,

```
$go.exe my_input.dat my_output.dat
```

where `go.exe` is the executable which reads data from file `my_input.dat` and, after processing, writes the output to file `my_output.dat`. This can be accomplished in Java via the `args[]` array to `main()`: `args[0]` holds the string `"my_input.dat"` and `args[1]` `"my_output.dat"`.

However we may opt for taking advantage of Java's graphical user interface (GUI) by instantiating an instance of `FileDialog`,

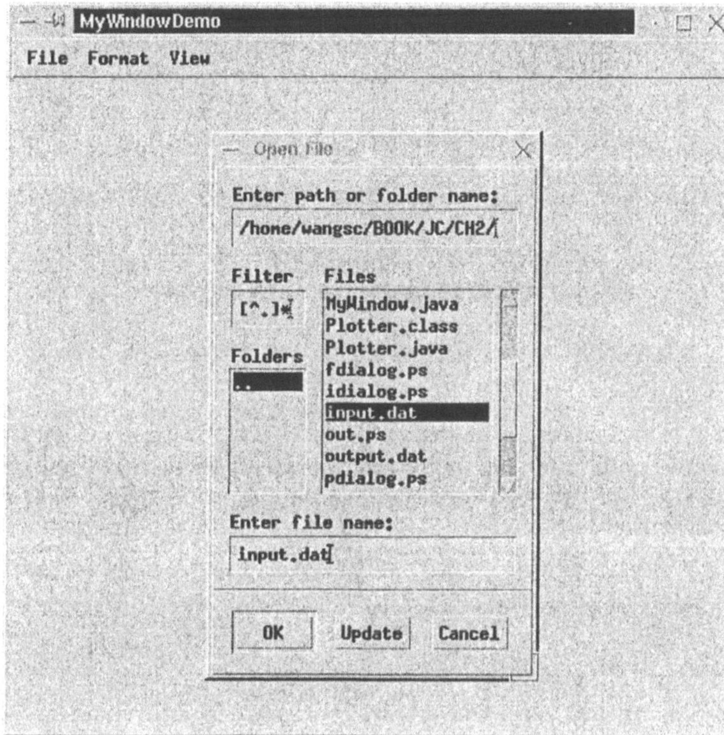

Figure 2.4. The file dialog

```
FileDialog opendlg = new FileDialog(this,"Open File", FileDialog.LOAD);
```

A dialog is shown upon user request, which is identified and acted upon in method `actionPerformed()`,

```
opendlg.show();
```

A screen shot of the dialog box is displayed in Figure 2.4. Object `opendlg`'s methods are then deployed to locate the file of user's choice where the mouse is released,

```
String infile = opendlg.getDirectory() + opendlg.getFile();
```

Before importing the data, we may want to specify the format of the data file. Listing 2.2 shows the raw data for the curves in Figure 2.1. In this example, three pieces of information can be supplied to the program: the total number of columns in the file, the column for the x coordinate data, and the number of lines to skip in the beginning of the input file. They are represented by the three integers `nColns`, `xIndex`, and `nSkips` in `MyWindow`. The interaction

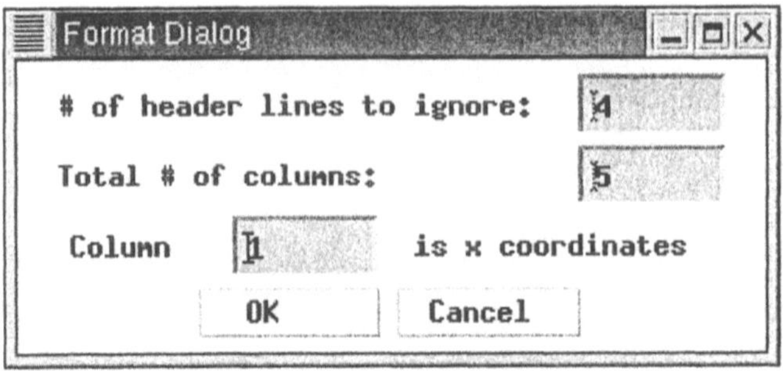

Figure 2.5. Dialog box for the user to update the input file format

medium between the user and the program is the dialog box in Figure 2.5. The class (`FDialog.java` in the appendix) also implements an `ActionListener` interface to read in user's input.

```
c
c these header lines are for comments
c data follow

1.10E+3 0.00879531 0.00237217 0.000756203 0.0
1.20E+3 0.0304673  0.0168901  0.0119284   0.051
1.25E+3 0.0863443  0.0337259  0.02265734  0.11
1.30E+3 0.251472   0.106202   0.05121746  0.278
1.35E+3 0.643502   0.306731   0.163953    0.611
1.40E+3 0.934565   0.713828   0.497001    0.938
1.45E+3 0.995856   0.962992   0.881213    0.985
1.50E+3 0.999473   0.997534   0.990032    0.987
1.55E+3 0.999735   0.999733   0.998151    0.991
```

Listing 2.2 Raw data in the input file for the curves in Figure 2.1

2.8 StreamTokenizer

To read numbers (or characters) from the input file, we introduce the versatile class `StreamTokenizer`, which appears in our method `loadData()` in `MyWindow.java`. A `try-catch` block is needed because the constructors and some methods of the first three classes in `loadData()` throw various exceptions.

First of all, the input file is wrapped into `FileInputStream` which creates a stream for reading data from a file,

```
FileInputStream fis  = new FileInputStream(infile);
```

`fis` is then wrapped into an instance of the class `InputStreamReader` which reads bytes and translates them into characters according to platform (or user) specified encodings,

```
InputStreamReader br = new InputStreamReader(fis);
```

The character-input stream is next buffered for efficient reading,

```
BufferedReader re     = new BufferedReader(br);
```

Finally, `re` is taken by the `StreamTokenizer` class and is parsed into tokens which are read once at a time,

```
StreamTokenizer sto  = new StreamTokenizer(re);
```

The first `while` block determines the total number of lines in the opened file. Arrays of appropriate size are then instantiated to store the data. The file is then closed by `FileInputStream`'s method `close()`. Objects `InputStreamReader`, `BufferedReader`, and `StreamTokenizer` which are created within the `try-catch` block expire after execution leaves the block and will be garbage collected when the system is free to do so.

The next `while` block opens again the file and the method `readNumber()` is now repeated `nLines - nSkips` times to fill the arrays in the `for` loop. The method makes use of methods of the class `StreamTokenizer`. `readNumber()` is able to read real numbers and numbers in scientific notation like 6.626E-34 and 1.37e10. The former is Planck's constant in $J \cdot s$ and the latter the age of the universe in years.

```
1100 0.0088 0.0024 0.0008 0.0000
1200 0.0305 0.0169 0.0119 0.0510
1250 0.0863 0.0337 0.0227 0.1100
1300 0.2515 0.1062 0.0512 0.2780
1350 0.6435 0.3067 0.1640 0.6110
1400 0.9346 0.7138 0.4970 0.9380
1450 0.9959 0.9630 0.8812 0.9850
1500 0.9995 0.9975 0.9900 0.9870
1550 0.9997 0.9997 0.9982 0.9910
```

Listing 2.3 Content of the saved file.

Most often, after data processing, we want to save the manipulated data. The file name for the output file can be entered by the user and then captured by the program in the way input files are opened for reading. The dialog box for this purpose is invoked by selecting the `Import` item in the `Format` menu and is shown in Figure 2.6. Here we demonstrate by writing the raw data of Listing 2.2 to an output file with the decimal format defined in the method `writeData()`. Listing 2.3 shows the content of the output file. They are seen to be the same as the raw data of Listing 2.2 except the numeral format.

During reading, saving and even drawing data, it is helpful to show a small message box on the screen, informing the user that work is in progress. Instances of such a class `Message` are created before and closed after the task. The type of the task being performed is specified as a string argument to the constructor of the class as shown in the example in Figure 2.7. The source code for this message class is listed in the appendix.

2.9 Graphics

We have so far annotated the class `MyWindow`, which sets up an interactive window environment for the class `Plotter`, whose job is to plot the input data.

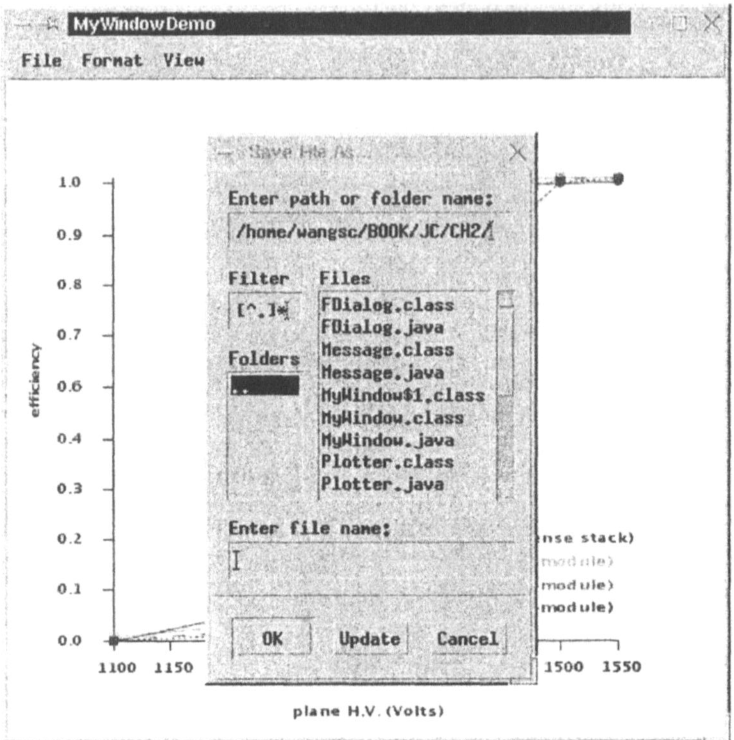

Figure 2.6. The dialog box for writing to files

Listing 2.4 shows the code of the plotting class. It is seen that it is subclassed from class `Canvas`, which is in turn a subclass of the class `Component`. The `Component` class, inheriting `java.lang.Object` which is the root of all Java classes, is the abstract superclass of many window classes. The class `Canvas` represents a blank rectangle on which graphics can be drawn and user input events are listened to. Unlike its parent `Component`, which is abstract, the class `Canvas` requires that its method `paint(graphics g)` be overridden by the programmer for customized graphics on the canvas. We now focus on the the method `paint()`.

```
/*      Sun-Chong Wang
        TRIUMF
        4004 Wesbrook Mall
        Vancouver, V6T 2A3
        Canada
        e-mail: wangsc@triumf.ca

        Plotter.java connects data points with color lines */

import java.lang.*;
import java.awt.*;
import java.awt.event.*;
import java.awt.geom.*;
import java.awt.font.*;
import java.text.DecimalFormat;
```

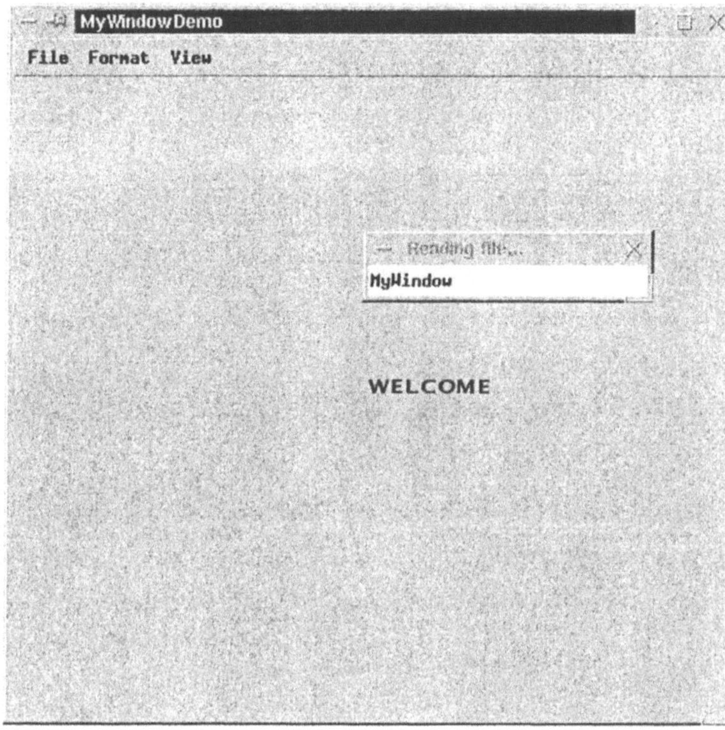

Figure 2.7. The message box showing that reading is underway

```
import java.util.*;

class Plotter extends Canvas {

 MyWindow parent;

 double xmin,xmax,ymin,ymax;
 double topborder,sideborder;
 static int bottom,right;
 int rectWidth = 6, rectHeight = 6;

 double x[];
 double y[][];

 final static int maxCharHeight = 20;
 final static int minFontSize = 8;

 public Plotter(MyWindow parent) {
        super();
        this.parent = parent;
 }

 public void paint(Graphics g) {
        final Color bg = Color.white;
        final Color fg = Color.black;
        final Color red = Color.red;
        final Color white = Color.white;
```

```
        final BasicStroke stroke = new BasicStroke(1.0f);
        final BasicStroke wideStroke = new BasicStroke(8.0f);

        final float dash1[] = {10.0f};
        final BasicStroke dashed1 = new BasicStroke(1.0f,
             BasicStroke.CAP_BUTT,BasicStroke.JOIN_MITER,10.0f,
             dash1, 0.0f);
        final float dash2[] = {2.0f};
        final BasicStroke dashed2 = new BasicStroke(1.0f,
             BasicStroke.CAP_BUTT,BasicStroke.JOIN_MITER,10.0f,
             dash2, 0.0f);
        FontMetrics fontMetrics;

        Graphics2D g2 = (Graphics2D) g;

        g2.setRenderingHint(RenderingHints.KEY_ANTIALIASING,
                            RenderingHints.VALUE_ANTIALIAS_ON);
        Dimension d = getSize();

        fontMetrics = pickFont(g2, "WELCOME", d.width);

        int rectWidth = 6, rectHeight = 6;
        int j,x0,y0,x1,y1;

        // this handles resizing of the window
        if ((d.width != right) || (d.height != bottom)) {
           bottom = d.height;
           right = d.width;
           SetScreenSize(right,bottom);
        }

        if (parent.beforePlot == false) {
           setBackground(Color.white);
           SetPlottingLimits();
           SetBorderSize(0.15,0.15);

           fontMetrics = pickFont(g2, "Vth = 200 mV", (d.width/6));
           // now draw the axes
           DrawXAxis(g2);
           DrawYAxis(g2);
           putAxisTitles(g2,stroke,d,"efficiency",20,
                                 "plane H.V. (Volts)",-50);

           drawPieces(g2,x,y[0],stroke,stroke,Color.green,1);
           drawPieces(g2,x,y[1],stroke,dashed1,Color.red,2);
           drawPieces(g2,x,y[2],stroke,dashed2,Color.black,3);
           drawPieces(g2,x,y[3],stroke,stroke,Color.blue,4);

           putCaption(g2,"100 mV (dense stack)",1400.,0.2,
                                            Color.blue,4);
           putCaption(g2,"200 mV (3-module)",1400.,0.15,
                                         Color.green,1);
           putCaption(g2,"350 mV (3-module)",1400.,0.10,Color.red,2);
           putCaption(g2,"500 mV (3-module)",1400.,0.05,
                                         Color.black,3);

        } else g2.drawString("WELCOME", d.width/2, d.height/2);
}       // end of re-display method

private void putCaption(Graphics2D g2, String text, double x,
                        double y, Color color, int symbol) {
        int i0, j0;
        g2.setPaint(color);
        i0 = GetXCoordinate(x);
        j0 = GetYCoordinate(y);
```

```
        i0 -= rectWidth/2;
        j0 -= rectHeight/2;
        switch (symbol) {
           case 1: g2.draw(new Rectangle2D.Double(i0-10, j0-5,
                                     rectWidth, rectHeight));
                   break;
           case 2: g2.draw(new Ellipse2D.Double(i0-10, j0-5,
                                     rectWidth, rectHeight));
                   break;
           case 3: g2.fill(new Rectangle2D.Double(i0-10, j0-5,
                                     rectWidth, rectHeight));
                   break;
           case 4: g2.fill(new Ellipse2D.Double(i0-10, j0-5,
                                     rectWidth, rectHeight));
           default:
        } // end of switch
        g2.drawString(text, i0, j0);
}

private void putAxisTitles(Graphics2D g2, BasicStroke stroke,
                     Dimension d, String xTitle, int yoffset,
                              String yTitle, int xoffset) {
        g2.setPaint(Color.black);
        g2.setStroke(stroke);
        g2.rotate(Math.toRadians(-90));
        g2.translate(-d.width/2,d.height/20);
        g2.drawString(xTitle,yoffset,0);

        g2.translate(d.width/2,-d.height/20);
        g2.rotate(Math.toRadians(90));
        g2.drawString(yTitle,d.width/2 + xoffset,d.height*29/30);
}

private void drawPieces(Graphics2D g2, double[] x, double[] y,
                   BasicStroke stroke, BasicStroke dashed,
                   Color color, int symbol) {
        int i0, j0;

        g2.setStroke(stroke);
        GeneralPath brokenLine = new GeneralPath(
                              GeneralPath.WIND_EVEN_ODD,y.length);
        brokenLine.moveTo(GetXCoordinate(x[0]), GetYCoordinate(y[0]));
        if (y.length > 0) {
           g2.setPaint(color);
           for(int j=0;j<y.length;j++) {
              i0 = GetXCoordinate(x[j]);
              j0 = GetYCoordinate(y[j]);
              if (j != 0) brokenLine.lineTo(i0, j0);
              i0 -= rectWidth/2;
              j0 -= rectHeight/2;
              switch (symbol) {
                  case 1: g2.draw(new Rectangle2D.Double(i0, j0,
                                     rectWidth, rectHeight));
                          break;
                  case 2: g2.draw(new Ellipse2D.Double(i0, j0,
                                     rectWidth, rectHeight));
                          break;
                  case 3: g2.fill(new Rectangle2D.Double(i0, j0,
                                     rectWidth, rectHeight));
                          break;
                  case 4: g2.fill(new Ellipse2D.Double(i0, j0,
                                     rectWidth, rectHeight));
                  default:
              }  // end of switch
```

```
            }
            g2.setStroke(dashed);
            g2.draw(brokenLine);
        }
}

// FontMetrics class encapsulates information on
// rendering a particular font on a particular screen
FontMetrics pickFont(Graphics2D g2, String longString, int xSpace) {
        boolean fontFits = false;
        Font font = g2.getFont();
        FontMetrics fontMetrics = g2.getFontMetrics();
        int size = font.getSize();
        String name = font.getName();
        int style = font.getStyle();

        while (!fontFits) {
           if ((fontMetrics.getHeight() <= maxCharHeight)
              && (fontMetrics.stringWidth(longString) <= xSpace)) {
              fontFits = true;
           } else {
              if (size <= minFontSize) {
                 fontFits = true;
              } else {
                 g2.setFont(font = new Font(name,style,--size));
                 fontMetrics = g2.getFontMetrics();
              }
           }
        }
        return fontMetrics;
}

public void DrawXAxis(Graphics2D g2) {
        int x0,x1,yofaxis,yoftick,yText;
        double xTickInterval,dValue;
        int tshift=11;        // number position offset

        x0 = (int) (sideborder*right);
        x1 = (int) ((1.0-sideborder)*right);
        if (ymin < 0) yofaxis = (int) ((1.0-topborder)*bottom +
                     (1.0-(2*topborder))*bottom*ymin/(ymax-ymin));
        else yofaxis = (int) ((1.0-topborder)*bottom);
        g2.draw(new Line2D.Double(x0, yofaxis, x1, yofaxis));

        xTickInterval = FindTicks(xmin,xmax);
        yoftick = yofaxis + (int) (topborder*bottom/10);
        yText = yofaxis + (int) (topborder*bottom/3);
        dValue = xmin;
        while (dValue <= xmax) {
              x0 = (int) (right*((1-2.*sideborder)*
                   (dValue-xmin)/(xmax - xmin))+right*sideborder);
              g2.draw(new Line2D.Double(x0, yofaxis, x0, yoftick));
              String fs = "0";
              DecimalFormat df = new DecimalFormat(fs);
              String sz = df.format(dValue);
              g2.drawString(sz, x0-tshift, yText);
              dValue+= xTickInterval;
        }
}

public void DrawYAxis(Graphics2D g2) {
        int y0,y1,xofyaxis,xoftick,xText;
        double yTickInterval,dValue;

        yTickInterval = FindTicks(ymin,ymax);
        y0 = (int) (topborder*bottom);
```

```
        y1 = (int) ((1-topborder)*bottom);

        if (xmin < 0) xofyaxis = (int)((1.0-2*sideborder)*
                     right*(-xmin/(xmax-xmin))+sideborder*right);
        else xofyaxis = (int) (sideborder*right);

        g2.draw(new Line2D.Double(xofyaxis, y0, xofyaxis, y1));
//        xText = (int) ((double) xofyaxis-(sideborder*right/4));
        xText = 5;
        xoftick = xofyaxis-(int) (sideborder*right/10);
        dValue = ymin;

        while (dValue<= ymax) {
              y0  = (int) ((1-topborder)*bottom-(1.0-2*topborder)*
                          bottom*(dValue-ymin)/(ymax-ymin));
              g2.draw(new Line2D.Double(xofyaxis, y0, xoftick, y0));
              String fs = "0.0";
              DecimalFormat df = new DecimalFormat(fs);
              String sz = df.format(dValue);
              g2.drawString(sz,xText+30,y0+3);
              dValue+= yTickInterval;
        }
}

public int GetYCoordinate(double dValue) {
        int y = (int) ((1-topborder)*bottom-(1.0-2*topborder)*
                     bottom*(dValue-ymin)/(ymax-ymin));
        return y;
}

public int GetXCoordinate(double dValue) {
        int x = (int) (right*((1-2.*sideborder)*(dValue-xmin)/
                     (xmax - xmin))+right*sideborder);
        return x;
}

public void SetScreenSize(int x, int y) {
        right = x;
        bottom = y;
}

public void SetPlottingLimits() {
        if (parent.beforePlot == false) {
           if (x.length > 0 && y[0].length > 0) {
              if ((GetXMin() == GetXMax()) ||
                  (GetYMin() == GetYMax())) ;
              else {
                    xmin = GetXMin();
                    xmax = GetXMax();
                    ymin = GetYMin();
                    ymax = GetYMax();
              }
           }
        }
}

public void SetBorderSize(double fraction_of_x,
                          double fraction_of_y) {
        if ((fraction_of_x <= 0) || (fraction_of_y <= 0)) ;
        else {
              topborder = fraction_of_y;
              sideborder = fraction_of_x;
        }
}

private double FindTicks(double AxisMin, double AxisMax) {
```

```
        double fSpan = 0;
        double multiplier = 1;
        double span,fInitialSpan;
        long lSpan,quot,rem;
        span = AxisMax - AxisMin;
        boolean b;

        if (AxisMax <= AxisMin)
           System.out.println("Error in axis data range");
        fInitialSpan = span;
        if (fInitialSpan < 10.0) {
           while (span < 10) {
                 multiplier *= 10;
                 span *= 10;
           }
        } else {
           while (span > 1.0e9) {
                 multiplier /= 10;
                 span /= 10;
           }
        }
        lSpan = (long) span;
        b = false;
        for (int i=10; i>=2; i--) {
            quot = lSpan/i;
            rem = lSpan - quot*i;
            if (rem == 0) {
               fSpan = (double) quot;
               fSpan = fSpan/multiplier;
               b = true;
            }
            if (b == true) break;
        }
        // if all else fails
        if (b == false)
        fSpan = (span/(2*multiplier)); // two intervals by default

        return fSpan;
}       // FindTicks method

public double GetXMin() {
        double dmin = x[0];
        for (int i=1; i<x.length; i++) if (x[i] < dmin) dmin = x[i];
        xmin = dmin;
        return xmin;
}

public double GetYMin() {
        double dmin = y[0][0];
        for (int j=0; j<y.length; j++) {
            for (int i=0; i<y[j].length; i++)
                if (y[j][i] < dmin) dmin = y[j][i];
        }
        ymin = dmin;
        return ymin;
}

public double GetXMax() {
        double dmax = x[0];
        for (int i=1; i<x.length; i++) if (x[i] > dmax) dmax = x[i];
        xmax = dmax;
        return xmax;
}

public double GetYMax() {
        double dmax = y[0][0];
        for (int j=0; j<y.length; j++) {
```

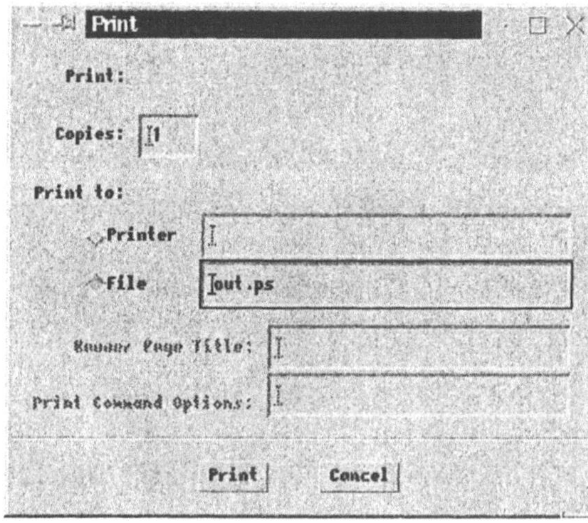

Figure 2.8. The printing dialog

```
            for (int i=0; i<y[j].length; i++)
                if (y[j][i] > dmax) dmax = y[j][i];
        }
        ymax = dmax;
        return ymax;
    }
}       // end of Plotter class
```

Listing 2.4 Plotter.java

First of all, color aliases are assigned to the variables of the `java.awt.Color` class, which is already imported. `black`, `blue`, `cyan`, `darkGray`, `gray`, `green`, `lightGray`, `magenta`, `orange`, `pink`, `red`, `white`, and `yellow` are defined in the variable field of `Color`. Other colors can also be created by the programmer. One such example is in the chapter of artificial neural network in part 2 of the book. The modifier `final` here means that those instances are made constant. `BasicStroke` defines how lines are drawn, i.e., solid, dotted, or dashed lines.

Functionality of the class `Graphics2D` is very rich in its own right. It is the fundamental class that renders 2-dimensional shapes, text, and images. It performs coordinate transformation and manages color, fonts, and text layout. It draws or fills circles, ovals, rectangles, polygons. Its sophistication is to fulfill demands of computer graphics and animation, which are topics of a whole book. We in this section grab what we need from `Graphics2D` to realize the screen shot that we saw in Figure 2.1. When tailoring for her drawing needs, the reader can leave most of the code intact except changing the axis captions, texts, and so forth.

`Graphics` is cast into `Graphics2D`,

```
Graphics2D g2 = (Graphics2D) g;
```

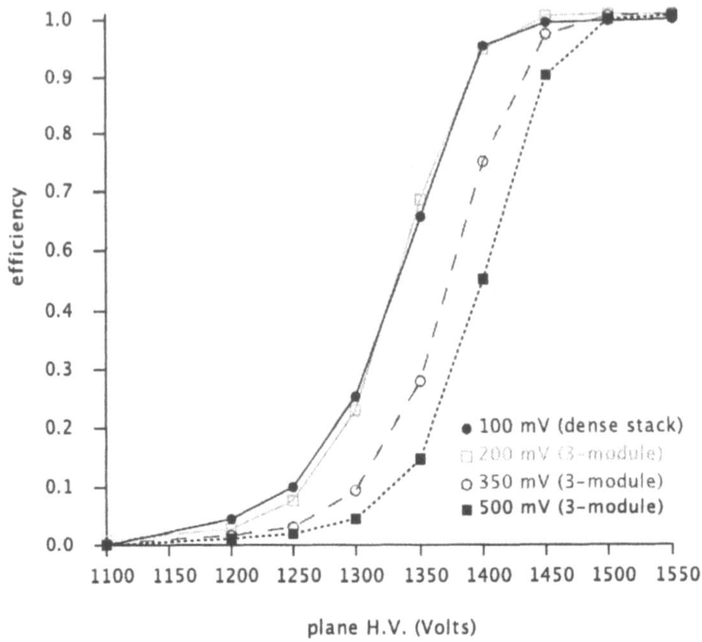

Figure 2.9. The graphics of the postscript output

Method `DrawXAxis()` draws horizontal and vertical axis, and particularly the ticks on the axes. Bounds of the axis are calculated by methods `getXMin()`, `getXMax()`, `getYMin()`, and `getYMax()`.

Data points read in `MyWindow` are first stored in instances of `GeneralPath`. A symbol (rectangle, or circle) is drawn on every point and then a line is drawn, connecting the data points. The line attribute (solid, dotted, ...) is set by `Graphics2D`'s `setStroke()` method. Note that data are converted to pixel coordinates whose origin is at the top-left corner of the screen by methods `GetXCoordinate()` and `GetYCoordinate()`. Texts are drawn by `Graphics2D`'s `drawString()` method at coordinates in the coordinate system of the data. Finally the method `pickFont()` picks the font of the right size such that the string "Vth = 200 mV" fits in the given restricted space.

When the user chooses the `Plot` item inside the `View` menu, the `paint()` method is called and drawing is engaged.

2.10 Printing

Recall that the class `MyWindow implements` the `Printable` interface in addition to `ActionListener`. We are therefore obliged to implement the `print()` method to render graphics to a printer. In short, when the `Print` option (Figure 2.3) is selected in the pull-down menu, a print dialog box of Figure 2.8 appears. Users of the application can choose to print out, to a printer or to a file (in postscript format), the graphics content in the canvas painted by `Plotter`'s `paint()` method. An example of the postscript output is shown in Figure 2.9.

2.11 Summary

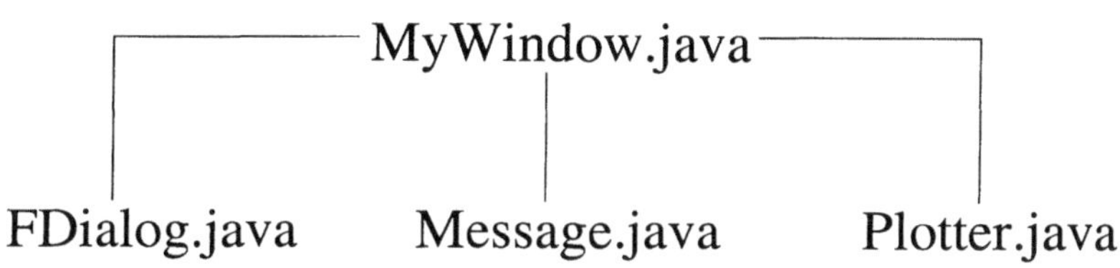

Figure 2.10. Source programs of the windowing/plotting/printing application

`FDialog.java` and `Message.java` are enclosed in the appendix.

Extensive graphics classes are provided in Java. The programmer's numerical objects can therefore couple with the graphics objects in a straightforward fashion, which manifests itself in various animation examples in later chapters of the book.

A pop-up dialog box serves as a convenient intermediary for user interaction with the application. Unlike parameter inputs from command-line arguments in traditional programs, users of Java applications can, through dialog boxes, choose data files or change parameter values relevant to the computational task. We will encounter more examples of parameter/outcome input/output through dialog boxes in the book.

Once an application implements the `Printable` interface, rendering can be on a printer or to the screen.

Interactivity of Java greatly enhances functionality of the language. Together with its portability, Java programming language shortens the time for collaborative software development and design.

2.12 References and Further Reading

K. Walrath and M. Campione, "The JFC Swing Tutorial: A Guide to Constructing GUIs (The Java(TM) Series)", Addison-Wesley Publishing Co. (1999)

D.M. Geary and A.L. McClellan, "Graphic Java", Prentice Hall, Englewood Cliffs, NJ (1997)

J. Knudsen, "Java 2D Graphics", 1st ed. O'Reilly & Associates, Inc. (1999)

Chapter 3

HIGH PERFORMANCE COMPUTING

Since the advent of the first digital computer decades ago, prices of computers have dropped significantly, making personal computers affordable. Meanwhile, the performances [in terms of the memory size and speed of the central processing unit (CPU)] double every 18 to 24 months (the so-called Moore's law). Computers are usually connected to one another to form a web of computers called Internet. A conceivable avenue of achieving high-performance computing is to coordinate together the vast number of otherwise idle computers on the network to tackle single tough tasks of computation. This is the very idea behind *grid computing* where both data and computing power are shared and accessible to a user. We will demonstrate an implementation of the so-called distributed computing to boost the performance in this chapter. Before this, let's introduce the other high performance computing via parallelism in Java.[1]

3.1 Parallel Computing

There arise cases where a task can be divided into independent pieces. If each piece is taken care of by an individual CPU and the multiple CPU's are run concurrently in the system, then ideally we expect a time saving by a factor of the total number of CPU's in the system. The actual saving depends on

[1]Language design features such as Java's checks of array indexes and references to objects for `out-of-bound` and `null`-pointer exceptions at runtime make Java a secure and reliable programming platform. They however have detrimental effects on technical computing. The *arrays of arrays* structure for multidimensional arrays in Java further hurts its numerical performance. As compiler optimization technologies advance, Java code can achieve 50% to 90% performance of highly optimized Fortran.

Since the book focuses on numerical computation, we play down handy applications of Java's container classes such as `java.util.Vector`. When collection classes are nevertheless used, we point out that the overhead due to extravagant object creation and type casting should be avoided.

the nature of the task and on the hardware architecture of the multi-processor system.

If, for example, data are shared among the processors, deliberate synchronization of the computing processes between data updates has to be devised. The issue of synchronization for jobs of a subtle nature like this is to be reminded of. Otherwise, the program ends up computing what is not meant to do because of corrupted data. We will see examples of synchronization in Chapter 13.

If the scattered computers are inter-connected via slow links, communications overhead counterplays gains in parallelization. The slower the link or the more frequent for the task to exchange data among computers, the severer is the penalty.

Bearing in mind the precautions, we show straightforward implementation of parallelism in Java, via the `Thread` class.

3.2 Java Threads

A thread is a separate executing process in a program. We have experienced, without notice, threads in Java's windowed programming in the previous chapter. Threading is in fact indispensable to interactivity in windowed programming. Consider the case where a lengthy job is running while the user clicks on a button on the menu of the window. Without threads, the application would not respond to the click until the lengthy job is finished. A window object is therefore a thread.

In a single CPU system, a thread is run by the CPU at one instance of time. It's suspended when, at the next instance of time, the CPU switches its attention to another thread. The switching over is usually so swift that the user does not notice the pause in the execution of individual threads. Examples are the playing of video/audio files while writing emails. In addition, consider that an application is faced with a slow process which can be due to hardware, such as reading from a slow device (tape) or network connection. It is then desirable for the application to spawn a separate thread for this slow process. The application can then impart time to other processes while waiting for the slow reading to finish its course. Chapter 13 shows one example of the so-called buffered I/O. Threading is thus handy for multi-process and/or multi-user applications. There, however, exist no gains in computation in such single CPU systems.

Once a class `extends` the `Thread` class (or `implements` the `Runnable` interface, a method called `run()` has to be provided. The method `run()` of the class then runs as a thread.

Dual-CPU PC's and quad-CPU servers are getting popularity due to their little extra cost. When they are administered by appropriate operating systems, such as GNU Linux, threads are dispatched to individual CPU's. It is this type of multi-processor systems that are gaining an edge with parallelization.

3.3 An Example of Parallel Computing

Consider the example of matrix multiplication,

$$L_{ij} = \sum_k M_{ik} N_{kj}. \tag{3.1}$$

Let's assume that i runs from 1 to 4 while j, k can presumably be very large. We further assume that there is a quad-processor system at our command. To speed up the process, we can split the multiplication of Eq. (3.1) into four pieces (threads) with processor one working on L_{1j} in thread one and processor two on L_{2j} in thread two, and so on. At the end of a thread, an array of size defined by the range of j is returned to the main program. We have to wait to make sure all the other threads are finished. The four returned arrays are then grouped into the matrix L before the program execution leaves the statement of Eq. (3.1). The benefit of the parallel computing in this example is appreciable when the size of the two multiplying matrices is large. We will see a real, similar implementation of parallel computation in Chapters 11 and 12.

3.4 Distributed Computing

Unlike parallel computing where divided jobs are loaded to multiple and usually identical processors to attain a speed-up in job execution, distributed computing usually relegates divided jobs to an echelon of disparate computers. Consider, for example, machine *S* has a very fast CPU while machine *C* is equipped with proprietary hardware (video card and monitor) for accelerated image processing and display. It is then advantageous to combine the merit of each, creating an improved system.

We describe how to achieve the goal with Java. The system architecture assumed is shown in Figure 3.1. The connection between the two machines is via the Internet (TCP/IP protocol). To set up such a distributed computing environment, the programmer is required to have a regular account on each machine; she does not have to be a superuser (system administrator) of either system.

3.5 Remote Method Invocation

Java provides an application package that enables a method to be run on a remote machine. The package is called Remote Method Invocation, or RMI for short. We envisage a scenario that a user sits in front of machine *C*, which hosts many of the user's utilities. In the application, however, a lengthy numerical calculation has to be undertaken before the result is further processed on

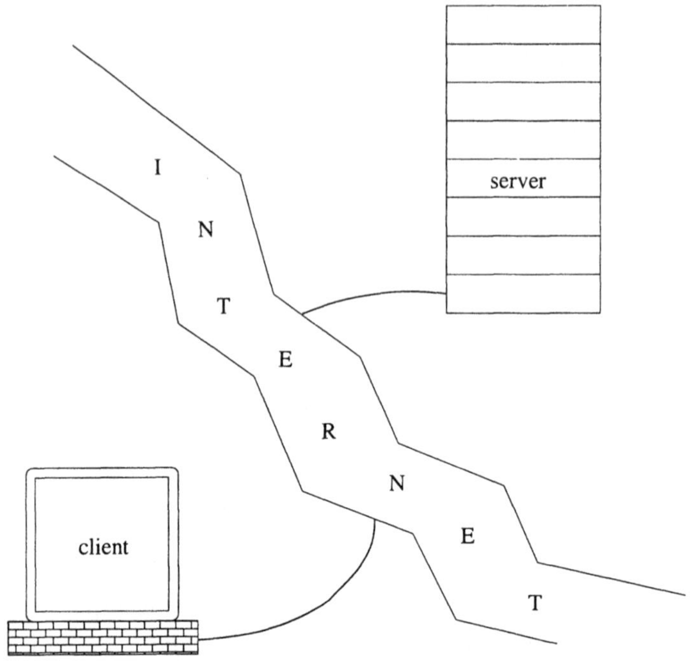

Figure 3.1. The presumed architecture for the distributed computing

machine C. To relieve the bottleneck, the calculation is to be sent to machine S, which is superior to C in terms of number crunching. In this case, machine S is referred to as RMI server and machine C RMI client.

Moreover, the content of the numerical calculation, implemented in a class, is up to the client; the server does not have to know the detail in advance. Namely, the class-loading onto machinc S is *dynamical*; the numerical class need not have to be precompiled and reside on the server. There are two ways of accomplishing dynamic loading of classes in Java. We can 'serialize' the class and then send it over the wire from the client to the server. The other is via a mechanism called *reflection*. The latter offers more flexibility regarding the content of the migrating objects and is the route we are taking. We coin RRMI for 'reflective RMI'.

3.6 An RMI Client

The code for the RRMI client is shown in Listing 3.1. At this point, the reader has hopefully gotten some familiarity with Java semantics. We are therefore going quickly through the code.

```
/*      Sun-Chong Wang
        TRIUMF
```

```
         4004 Wesbrook Mall
         Vancouver, V6T 2A3
         Canada
         e-mail: wangsc@triumf.ca

         RRMIClient.java is the client program of the
         'reflective remote method invocation' application */

package rrmiclient;

import java.rmi.*;
import java.math.*;
import java.io.*;
import java.net.*;

import rrmiinterf.*;

public class RRMIClient {
 public static void main(String args[]) {
         if (System.getSecurityManager() == null) {
            System.setSecurityManager(new RMISecurityManager());
         }

         int port = 2001;
         String classpath = "/home/wangsc/JAVA/RRMI";

         try {
             new ClassFileServer(port, classpath);
         } catch (IOException e) {
             System.out.println("Unable to start ClassServer: " +
                                  e.getMessage());
             e.printStackTrace();
         }

         try {
             URL url = new URL("http://lin06.triumf.ca:2001/");
             String name = "//" + args[0] + "/Compute";
             RRMIInterf comp = (RRMIInterf) Naming.lookup(name);

             Object Arg[] = new Object[1];
             Arg[0] = new Integer(10);      // 10 digits
             comp.rroc(url, "rrmiclient.MyDemo","MyDemoObj");
             BigDecimal pi = (BigDecimal)
                             comp.rrmi("MyDemoObj","computePi",Arg);
             System.out.println("PI = "+pi);
/*
             // the other way of invoking pi calculation
             comp.rroc(url, "rrmiclient.MyPi", "MyPiObj", Arg);
             BigDecimal pi1 = (BigDecimal)
                              comp.rrmi("MyPiObj","execute");
             // similarly
             Arg[0] = new Integer(20);      // reset to 20 digits
             BigDecimal pi2 = (BigDecimal)
                              comp.rrmi("MyPiObj","computePi",Arg);
             System.out.println("PI = "+pi1+" (or "+pi2+")");
*/
             Object Args[] = new Object[3];
             Args[0] = new String("Say_Hello");
             Args[1] = new Integer(999);
             Args[2] = new Double(1000.0);
             System.out.println(comp.rrmi("MyDemoObj","SetDataString",
                                                                 Args));
             Object arr[] = new Object[1];
```

```
            double[][] da = new double[3][3];
            da[0][0] = 1.0; da[0][1] = 2.0; da[0][2] = 3.0;
            arr[0] = da;
            double[][] arrcast = (double[][])
                       comp.rrmi("MyDemoObj", "SetDoubleArray",arr);
            System.out.println(arrcast[0][0]+" "+arrcast[0][1]+" "+
                                               arrcast[0][2]);
        } catch (Exception e) {
            System.err.println("RRMIClient exception: " +
                                          e.getMessage());
            e.printStackTrace();
        }
    }
}   // end of client class
```

Listing 3.1 RRMIClient.java

The class `RRMIClient` is really simple. It contains only the `main()` method, where firstly a security manager is set up. The `ClassFileServer` class is for transferring supporting classes which are, among other things, necessary for the RMI mechanism. `ClassFileServer` can be relinquished if the client machine (machine *C* in this case) runs an HTTP web server. Class `ClassFileSer ver` here therefore serves as a mini-HTTP server. It however makes no hindrance even if a web server does run in the client machine. Note that `port` and `classpath` are the two variables that the reader needs to change to suit her circumstances.

`URL` is the class holding location information of the client machine.

The command line argument `args[0]` carries the domain name of the server machine with which this client program/machine is to contact. After the server, together with the services registered on it, is looked up, the server returns to the client an `RRMIInterf` object. Note that, under the surface, a class called `stub` is also returned. Subsequent client calls to the server is then in fact through the `stub`. We focus here on a working example of RMI. Development of RMI is itself a specialized and evolving subject and beyond the scope of this text.

3.7 The Remote Interface

`RRMIInterf`, Listing 3.2, is an interface class which encapsulates the remote services the server offers. It `extends Remote`. Recall that we are sending (numerical) objects to the server. We firstly need to create the object on the server. This is done by invoking the remote method (in the client program),

```
rroc(URL url, String ClassName, String ObjName, Object[] Args);
```

where `url` tells the server the address of the client. `ClassName` is the name of the migrating class. `ObjName` specifies the object name of the migrating class when it is instantiated (in the server machine). `Args`, is optional and holds the arguments, when needed, for the constructor of the class. After this method, the class, which originally resides on the client machine, is created on the server machine. The mechanism is through reflection, as shown in the server code. The reflection mechanism is detailed in Section 3.10.

```
/*      Sun-Chong Wang
        TRIUMF
        4004 Wesbrook Mall
        Vancouver, V6T 2A3
        Canada
        e-mail: wangsc@triumf.ca
```

```
        RRMIInterf.java is the interface class for the
        'reflective remote method invocation' application.
        It tells what services the server provides */

package rrmiinterf;

import java.rmi.Remote;
import java.rmi.RemoteException;
import java.net.*;

public interface RRMIInterf extends Remote {
       void rroc(URL url, String ClassName, String ObjName)
                                         throws RemoteException;
       void rroc(URL url, String ClassName, String ObjName,
                             Object[] Args) throws RemoteException;
       Object rrmi(String ObjName, String MethodName)
                                         throws RemoteException;
       Object rrmi(String ObjName, String MethodName,
                             Object[] Args) throws RemoteException;
}
```

Listing 3.2 RRMIInterf.java

Next, the method of the newly created object on the server is invoked by the other service the server provides,

```
Object rrmi(String ObjName, String MethodName, Object[] Args);
```

where `ObjName` is the name of the object assigned to the class in the previous `rroc()` method. `MethodName` is the name of the method of the sent object the programmer intends to invoke. `Args`, which is optional, passes the arguments needed to invoke the method `MethodName()`.

Note that since `Object` is the root of all other classes, `Args` in `rroc` and `rrmi` can represent `String`'s, any of the primitive data types in Java, and their arrays. In the same Java virtual machine (JVM), argument passing is by value (or copy) for primitive data types and by reference (pointer or memory address) for objects. Across JVM's, as in the case of distributed computing, however, objects are copied and then passed to or returned from the method. The rest of the client program is simply exercising the use of `rroc` and `rrmi`, especially how parameters are prepared and passed to them. For example, a 2-dimensional array is represented by an `Object` object before passing, and then downcast to a 2-dimensional array after returned from the method call.

Note that, in the client program, the only reference to `MyDemo` (Listing 3.3) is in `rroc`: `"MyDemo"`, which is simply a string. Therefore when the client program is compiled, the compiler checks nothing about `MyDemo.java`. In fact, the programmer has to issue a separate command to compile `MyDemo.java`. The prescription in the interface `RMIInterf` therefore serves as the the first guard against sending ill-posed method calls to the server.

```
/*      Sun-Chong Wang
        TRIUMF
        4004 Wesbrook Mall
        Vancouver, V6T 2A3
        Canada
        e-mail: wangsc@triumf.ca

        MyDemo.java is one of the migrating objects to the
        'reflective remote method invocation' server where
        methods in this class will be run */
```

```
package rrmiclient;

import java.io.Serializable;
import java.math.*;

public class MyDemo implements Serializable {
 final int Dimx = 3; // dimension of the arry
 final int Dimy = 3;
 String s;
 int MyI;
 double[][] M;

 public MyDemo() {
        super();
        M = new double[Dimx][Dimy];
 }

 public double SetDataString(String set, int I, double d) {
        s = set;
        MyI = I;
        M[0][0] = d;
        return ((double) MyI+M[0][0]);
 }

 public Object SetDoubleArray(Object MM) {
        double[][] da = (double[][]) MM;
        if (da.length == Dimx && da[0].length == Dimy) {
           for (int i=0; i<Dimx; i++) {
               for (int j=0; j<Dimy; j++) {
                   M[i][j] = da[i][j];
               }   // j loop
           }   // i loop
           return M;
        } else { return null; }
 }

 public BigDecimal ComputePi(int digit) {
        MyPi demo = new MyPi();
        return demo.computePi(digit);
 }
}       // end of MyDemo class
```

Listing 3.3 MyDemo.java

Even though the class `MyPi` of Listing 3.4 does not appear in `RRMIClient.java`, it will still be automatically loaded to the server since it is referenced to in the class `MyDemo`.

```
/*      Sun-Chong Wang
        TRIUMF
        4004 Wesbrook Mall
        Vancouver, V6T 2A3
        Canada
        e-mail: wangsc@triumf.ca

        MyPi.java is adopted from the 'Pi.java' in Sun's
        RMI tutorial web site. This objects migrates from the
        RRMI client machine to the RRMI server machine where
        computation methods in this class are executed */

package rrmiclient;

import rrmiinterf.*;
import java.math.*;

public class MyPi {
```

```
/** constants used in pi computation */
private static final BigDecimal ZERO =
    BigDecimal.valueOf(0);
private static final BigDecimal  ONE =
    BigDecimal.valueOf(1);
private static final BigDecimal FOUR =
    BigDecimal.valueOf(4);

/** rounding mode to use during pi computation */
private static final int roundingMode =
    BigDecimal.ROUND_HALF_EVEN;

/** digits of precision after the decimal point */
private int digits;

/**
 * Construct a task to calculate pi to the specified
 * precision.
 */

public MyPi() {
}

public MyPi(int digits) {
    this.digits = digits;
}

public Object execute() {
    return computePi(digits);
}

/**
 * Compute the value of pi to the specified number of
 * digits after the decimal point.  The value is
 * computed using Machin's formula:
 *
 *          pi/4 = 4*arctan(1/5) - arctan(1/239)
 *
 * and a power series expansion of arctan(x) to
 * sufficient precision.
 */
public static BigDecimal computePi(int digits) {
    int scale = digits + 5;
    BigDecimal arctan1_5 = arctan(5, scale);
    BigDecimal arctan1_239 = arctan(239, scale);
    BigDecimal pi = arctan1_5.multiply(FOUR).subtract(
                         arctan1_239).multiply(FOUR);
    return pi.setScale(digits,
                       BigDecimal.ROUND_HALF_UP);
}
/**
 * Compute the value, in radians, of the arctangent of
 * the inverse of the supplied integer to the speficied
 * number of digits after the decimal point.  The value
 * is computed using the power series expansion for the
 * arc tangent:
 *
 * arctan(x) = x - (x^3)/3 + (x^5)/5 - (x^7)/7 +
 *     (x^9)/9 ...
 */
public static BigDecimal arctan(int inverseX,
                                int scale) {
    BigDecimal result, numer, term;
    BigDecimal invX = BigDecimal.valueOf(inverseX);
    BigDecimal invX2 =
        BigDecimal.valueOf(inverseX * inverseX);
```

```
        numer = ONE.divide(invX, scale, roundingMode);

        result = numer;
        int i = 1;
        do {
            numer =
                numer.divide(invX2, scale, roundingMode);
            int denom = 2 * i + 1;
            term =
                numer.divide(BigDecimal.valueOf(denom),
                             scale, roundingMode);
            if ((i % 2) != 0) {
                result = result.subtract(term);
            } else {
                result = result.add(term);
            }
            i++;
        } while (term.compareTo(ZERO) != 0);
        return result;
    }
}   // end of MyPi class
```

Listing 3.4 MyPi.java

3.8 Serialization

The class `MyPi` in Listing 3.4 implements `Serializable`. This interface serializes the (compiled) object, `MyPi`. Upon arriving at machine *S*, `MyPi` is recovered by automatic de-serialization. The compiled `MyPi` can then be run on machine *S*. In a far more sophisticated case, an RMI master server can manage a cluster of servers. The `Serializable` property makes it convenient when the master server decides, based on consideration of load balancing, to further relegate this object to other slave machines.

3.9 A Reflective RMI Server

We now turn to Listing 3.5 for the code of the server class `RRMIServer`. What it extends and implements amount to make it an RMI server. Note the interface `RRMIInterf` which serves as the contract between this server and the client. Also noted is the exception throwing, which is mandatory since machine *S* or network connections might go down.

```
/*      Sun-Chong Wang
        TRIUMF
        4004 Wesbrook Mall
        Vancouver, V6T 2A3
        Canada
        e-mail: wangsc@triumf.ca

        RRMIServer.java establishes the server for the
        'reflective remote method invocation' application */

package rrmiserver;

import java.rmi.*;
import java.rmi.server.*;
import java.io.*;
import java.lang.reflect.*;
import java.net.*;

import rrmiinterf.*;

public class RRMIServer extends UnicastRemoteObject
                        implements RRMIInterf {
```

```
final int max_o = 20;   // maximum no. of objects
Object objs[][];        // the object bank
int oindex;

public RRMIServer() throws RemoteException {
        super();
        objs = new Object[max_o][2];
        oindex = 0;
}

public void rroc(URL url, String ClassName, String ObjName) {
        boolean loaded = false;

        for (int i=0; i<max_o; i++) {
            if (ObjName.equals((String)objs[i][1])) {
               loaded = true;
               break;
            }
        }
        if (loaded == false) {
           try {
               Class cls = RMIClassLoader.loadClass(url, ClassName);
               objs[oindex%max_o][0] = cls.newInstance();
               objs[oindex%max_o][1] = ObjName;
               oindex += 1;
               System.out.println(oindex+" objects loaded");
           } catch (ClassNotFoundException cnfe) {
               System.out.println("class not found: "+
                                                cnfe.getMessage());
           } catch (Throwable e) {
               System.out.println("rroc Error: "+e.getMessage());
           }
        } // if loaded
}

public void rroc(URL url, String ClassName, String ObjName,
                                        Object[] Args) {
        int j = 0;
        boolean loaded = false;

        for (int i=0; i<max_o; i++) {
            if (ObjName.equals((String)objs[i][1])) {
               loaded = true;
               break;
            }
        }
        if (loaded == false) {
           try {
               Class cls = RMIClassLoader.loadClass(url, ClassName);
               Constructor[] ctlist = cls.getDeclaredConstructors();
               for (int i=0; i<ctlist.length; i++) {
                   Class pvec[] = ctlist[i].getParameterTypes();
                   if (pvec.length == Args.length) { j = i; break; }
               }
               objs[oindex%max_o][0] = ctlist[j].newInstance(Args);
               objs[oindex%max_o][1] = ObjName;
               oindex += 1;
               System.out.println(oindex+" objects loaded");
           } catch (ClassNotFoundException cnfe) {
               System.out.println("class not found: "+
                                                cnfe.getMessage());
           } catch (Throwable e) {
               System.out.println("rroc Error: "+e.getMessage());
           }
        } // if loaded
}
```

```
public Object rrmi(String ObjName, String MethodName) {
        int j1 = 0;
        int j2 = 0;

        try {
           for (int i=0; i<max_o; i++) {
               if (ObjName.equals((String)objs[i][1])) j1 = i;
           }
           Class cls = objs[j1][0].getClass();
           Method methlist[] = cls.getDeclaredMethods();
           for (int i=0; i<methlist.length; i++) {
               if (MethodName.equals(methlist[i].getName())) {
                  j2 = i;
                  break;
               }
           }
           Object[] Args = new Object[0];
           Object retobj = methlist[j2].invoke(objs[j1][0], Args);
           return retobj;
        } catch (Throwable e) {
           System.out.println("rrmi Error: "+e.getMessage());
           return null;
        }
}

public Object rrmi(String ObjName, String MethodName,
                                              Object[] Args) {
        int j1 = 0;
        int j2 = 0;

        try {
           for (int i=0; i<max_o; i++) {
               if (ObjName.equals((String)objs[i][1])) j1 = i;
           }
           Class cls = objs[j1][0].getClass();
           Method methlist[] = cls.getDeclaredMethods();
           for (int i=0; i<methlist.length; i++) {
               if (MethodName.equals(methlist[i].getName())) {
                  j2 = i;
                  break;
               }
           }
           Object retobj = methlist[j2].invoke(objs[j1][0], Args);
           return retobj;
        } catch (Throwable e) {
           System.out.println("rrmi Error: "+e.getMessage());
           return null;
        }
}

public static void main(String[] args) {
        if (System.getSecurityManager() == null) {
           System.setSecurityManager(new RRMISecurityManager());
        }

        int port = 2001;
        String classpath = "/home5/wangsc/JAVA/RRMI";
        try {
            new ClassFileServer(port, classpath);
        } catch (IOException e) {
            System.out.println("Unable to start ClassServer: " +
                                    e.getMessage());
            e.printStackTrace();
        }
```

```
        String name = "//lin01.triumf.ca/Compute";
        try {
            RRMIInterf engine = new RRMIServer();
            Naming.rebind(name, engine);
            System.out.println("RRMIEngine bound");
        } catch (Exception e) {
            System.err.println("RRMIEngine exception: " +
                                            e.getMessage());
            e.printStackTrace();
        }
    }   // end of main
}       // end of server class
```

Listing 3.5 RRMIServer.java

Let's firstly look at the `main()` method of the server. Again, the `ClassFile Server` is present when an HTTP web server is not available on the server machine.

The string `//lin01.triumf.ca/Compute` registers the service available on this server and is to be looked up by the clients. The method `Naming.rebi nd()` binds the symbol of registration to the implementation. After this point, the server is set up.

3.10 Reflection

We now look at the implementation of the services. In the method `rroc()`, first of all, if `ObjName` is not matched to any name of the already existing objects in the object bank on the server, an instance of the object is created. To create, we first call the method `RMIClassLoader.loadClass()` to load the class from the client. We then invoke the constructor whose number of parameters matches that of the passing array: `Args`. An instance of the object is then stored in the object bank.

Now when the method `rrmi()` is invoked by the client, the object represented by the passing parameter `ObjName` is retrieved from the object bank. The method that is to be invoked by the client is searched for from the available methods of the object. It is then invoked by the `invoke()` method of the class `Method`.

Introspecting upon the class itself, finding out about the constructors, methods, ..., among others, are facilitated by the reflection mechanism in Java. All the server needs to run the sent classes are the names of the classes and the names of the methods *at runtime !*

3.11 Build and Run the Server

On the server machine, suppose the directory the programmer is working is `/home5/wangsc/JAVA/RRMI` (called working directory hereafter). Now create `rrmiinterf` directory under this working directory and put the file `RRMIInterf.java` into this subdirectory. Again under the working directory, create `rrmiserver` directory, under which will reside files `ClassFileServer .java`, `ClassServer.java`, `RRMISecurityManager.java`, and `RRMIServ er.java`. Class `ClassServer` is the superclass of `ClassFileServer`. Class `RRMISecurityManager`, Listing 3.6, simply extends the default `RRMISecuri tyManager` class for security management.

```
/*      Sun-Chong Wang
        TRIUMF
```

```
        4004 Wesbrook Mall
        Vancouver, V6T 2A3
        Canada
        e-mail: wangsc@triumf.ca

        RRMISecurityManager.java is the security manager of the
        'reflective remote method invocation' application */

package rrmiserver;

import java.rmi.RMISecurityManager;

public class RRMISecurityManager extends RMISecurityManager {
    public void checkMemberAccess(Class clazz, int which) {
        // void
    }
    public void checkPackageAccess(String pkg) {
            super.checkPackageAccess(pkg);
        // void
    }
}
```

Listing 3.6 RRMISecurityManager.java

Now set the class path to the working directory. In a bash shell of GNU Linux, this is done by ($ stands for the system prompt),

```
$export CLASSPATH=/home5/wangsc/JAVA/RRMI
```

The csh shell equivalent is,

```
$setenv CLASSPATH /home/wangsc/JAVA/RRMI
```

Now under the working directory, issue (Note all the following commands are issued from the working directory.),

```
$javac rrmiinterf/RRMIInterf.java
```

Next, do,

```
$javac rrmiserver/RRMIServer.java
```

and then,

```
$rmic -d .  rrmiserver.RRMIServer
```

Now set the class path to none by, for example,

```
$export CLASSPATH=
```

in the bash shell. Now, issue,

```
$rmiregistry &
```

Then set the class path back to the working directory,

```
$export CLASSPATH=/home5/wangsc/JAVA/RRMI
```

Finally, issue,

```
$java -Djava.rmi.server.codebase=http://lin01.triumf.ca:200
```

```
1/ -Djava.security.policy=java.policy rrmiserver.RRMIServe
r
```

where `lin01.triumf.ca` is the name of the server machine and file `java.policy` (in chapter appendix) grants permissions for class file transfers. You then see,

```
$RRMIEngine bound
```

on the screen. The server is now on, ready for jobs from the client !

The subdirectories we put various source programs and the working directory we issue compilation commands result from the `package` statement we defined in the beginning of each program.

3.12 Build and Run the Client

Building the client is much simpler. Suppose the working directory on the client is `/home/wangsc/JAVA/RRMI`. We need the interface class. So create the `rrmiinterf` subdirectory under the working directory and then transfer `RRMIInterf.class` from the server machine to this subdirectory.

Now, again, set the class path to the working directory. Then compile the client source,

```
$javac rrmiclient/RRMIClient.java
```

Then compile the two migrating object sources,

```
$javac rrmiclient/MyPi.java
$javac rrmiclient/MyDemo.java
```

The client is built. Now run it by,

```
$java -Djava.security.policy=java.policy rrmiclient.RRMICli
 ent lin01.triumf.ca
```

You will get,

```
$reading:  rrmiclient.MyDemo
$reading:  rrmiclient.MyPi
$PI = 3.1415926536
$1999.0
$1.0 2.0 3.0
```

on the screen of the client machine, and,

```
$reading:  rrmiserver.RRMIServer_Stub
$1 objects loaded
```

on the screen of the server machine. Congratulations ! A distributed computing environment via reflection and RMI is successfully set up.

Because of the heterogeneity of the computers on the Web and the variability of network performance, more fault tolerance is desirable for a better server. Furthermore, to optimize usage of resources, a manager server might want to relay sent objects to other less loaded machines. On the client side, after requesting a remote service, the client might want to work on other objects instead of keeping waiting for server's replies. This concurrency can be accomplished by Java threads.

3.13 Summary

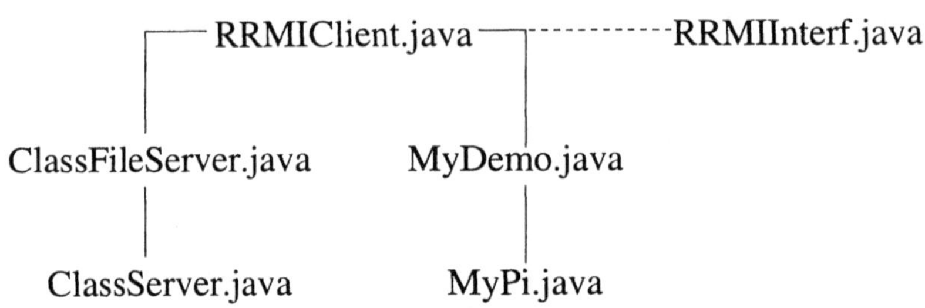

Figure 3.2. Source programs for the client of the reflective RMI

Supporting classes `ClassFileServer.java` and `ClassServer.java` can be found in Sun Microsystems' Java website. `java.policy` is enclosed in the appendix of this chapter.

We took advantage of Java's utilities of Remote Method Invocation and reflection to demonstrate distributed computing in Java. In this application, classes migrate from the client to the server at runtime. Objects of the migrating classes are instantiated on the sever. Whenever the numerical methods are called in the client application, computation is run on the server machine and results automatically returned to the client application.

Parallelization is suitable for some particular numerical applications, such as the genetic algorithm and path integrals to be introduced later in the book.

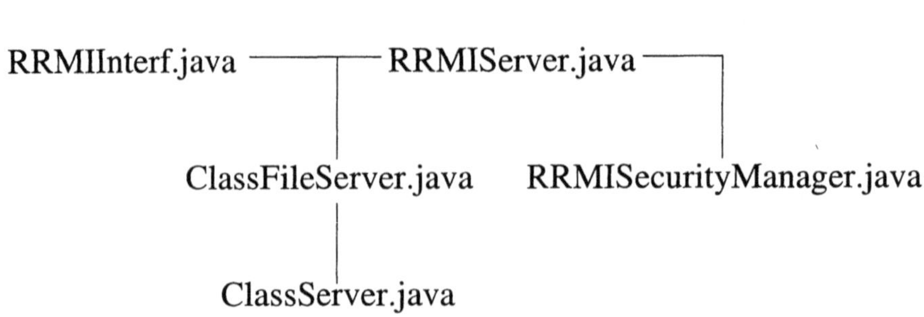

Figure 3.3. Source programs for the server of the the reflective RMI

Hardware-wise, the programmer has to carefully assess effects of data sharing and transferring among processors. In contrast, by extending the `Thread` class, parallel computing in Java is relatively straightforward (except the synchronization issue we mentioned earlier).

3.14 Appendix

```
grant {
    permission java.net.SocketPermission "*:1024-65535",
        "connect,accept";
};
```

Listing 3.7 java.policy

3.15 References and Further Reading

Search Sun Microsystems' Java website at `java.sun.com` for RMI tutorials

J. Farley, "Java Distributed Computing", 1st ed. O'Reilly & Associates, Inc. (1998)

S. Oaks and H. Wong, "Java Threads", 2nd ed. O'Reilly & Associates, Inc. (1999)

S. Oaks, "Java Security", 2nd ed. O'Reilly & Associates, Inc. (2001)

PART II

COMPUTING

Chapter 4

SIMULATED ANNEALING

In many applications, the parameters which are sought to minimize (or maximize) the objective function are not continuously varying. For instance, a salesperson is to travel through a series of cities in an order that gives the shortest traveling distance. The parameters take the form of integers in this case. And as the number of cities increases, the number of possible configurations (sequences) grows rapidly, rendering exhaustive search infeasible.

Conventional greedy minimization methods, such as the gradient method to be introduced in Chapter 11, might also fail in cases where there exist many local minima in the configuration space. Preempting an imminent gain may fend one off the final maximum payoff. In this chapter, we show an algorithm that emulates the way nature works in search for global extrema.

4.1 Introduction

At high temperatures, the atoms (or molecules) of a liquid move around freely, colliding with one another. As the liquid cools (by exchanging energy with an external reservoir), atoms lose their mobility. If the cooling takes place slowly enough, the atoms find their best mutual positions and the liquid finally crystallizes. The resulting crystalline structure exhibits uniformity that can extend over a distance that is an indefinite multiple of the size of the individual atom. Since the crystal is stable its energy is the lowest. The slowly cooling down process is called *annealing*. If, on the other hand, the temperature dropping is of a sudden, a solid still forms but the resulting structure may contain domains and the solid is prone to fracture along the domain boundaries. In the case of quenching, the total energy of the atomic system settles into a local, instead of global, minimum.

Applying annealing to the traveling salesman problem, we, starting from a trial order of traveling, shuffle the order (which can be stored in an array) and then calculate the total traveling distance. If the new order gives a shorter distance, it is accepted and serves as the current best order from which next shuffling starts. Otherwise, the trial order is discarded. As the 'temperature' drops, the shuffling gets gentler. A gentler shuffling can be implemented by, for example, picking up a smaller number of cities to be swapped. The process continues until there is no improvement.

The annealing lesson tells us of cooling slowly in the search of the global minimum. We still lack of the key ingredient which dispenses us from the lure of local minima. Furthermore, it is helpful to quantify 'high', 'low' temperatures, and the 'slowness' of the cooling. The former is achieved by the Metropolis algorithm. A thermodynamics analysis of the system gives us insights to the latter.

4.2 Metropolis Algorithm

Energy minimizing drives a system toward its destination. However, a second, equally important factor comes in to compete. That is *entropy maximizing*. Entropy is a measure of randomness of a system, and is proportional to the temperature of the system. A system in equilibrium at temperature T has a relative chance, $P(E)$, of staying at a state of energy E,

$$P(E) \propto \exp(-\frac{E}{k_B T}), \tag{4.1}$$

where k_B, Boltzmann's constant, relates temperature T to energy E. A system therefore tends to stay in a state of energy as low as allowed by the temperature. The probability, P_{tr}, for the system to transit from a state of low energy E_l to a state of high energy E_h is, according to Eq. (4.1),

$$P_{tr}(E_l \to E_h) = \exp(-\frac{E_h - E_l}{k_B T}). \tag{4.2}$$

This nonvanishing transition probability, albeit small, bails us out of local minima, and the method successfully implementing the transition probability is called Metropolis algorithm.[1]

Simulated annealing is thus a powerful method of optimization; it is a temperature cooling procedure embedding Metropolis algorithm. Yet the idea is simple, implementations leading to efficient searches depend on particular problems at hand and often require some experimentation.

[1] We will encounter in Chapter 10 an example of Metropolis algorithm in a different context.

4.3 Ising Model

To manifest the annealing process, we consider the Ising model as an idealized example. Similar to the traveling salesman problem, the parameters in an Ising system are discrete. Moreover, they take only two values: either +1 or -1. Ferromagnetism is the property of a material which becomes magnetic when its temperature is below some critical value. The ferromagnetism can be modeled by a simple picture where each atom of the material carries a parameter called *spin* (magnetic moment), which can point either up or down. When the spins of the composing atoms are randomly oriented, as at high temperatures, the total spin of the system averages zero and the material is non-magnetic. As the temperature is lowered, the spins of individual atoms in domains start aligning with one another, and the material can become slightly magnetic. As the temperature drops further, the material becomes more and more magnetic until at the lowest temperature all the spins line up (or down). The model is reminiscent of the freezing of water. As the temperature drops across zero centigrade (the critical temperature), water crystallizes into ice.

Near the critical temperature T_c, the magnetization (average spins), M, of the ferromagnet varies with temperature T as,

$$M = \begin{cases} 0 & \text{for } T > T_c \\ M_0 \left(\frac{T_c - T}{T_c}\right)^{\beta} & \text{for } T \leq T_c \end{cases} \tag{4.3}$$

where M_0 is a proportional constant and β is called *critical exponent*, whose value depends only on the dimensionality and symmetry of the system, and not on other details of the system. For example, $\beta = 0.125$ (≈ 0.33) for 2 (3)-dimensional Ising systems. Therefore, the Ising model and its variants, despite the simplicity, are, based on the property of universality, believed to represent a whole large set of other, more complicated systems in nature. It is, therefore, one of the most studied systems in statistical physics. For 3-dimensional Ising models, there have, however, existed no analytical solutions, and many calculations are based on computer simulations.

The energy (or called Hamiltonian) of an Ising system is represented as follows,

$$E = -\frac{J}{2} \sum_{<i,j>} s_i s_j, \tag{4.4}$$

where s_i = +1 or -1 is the spin of the atom, $< i, j >$ means summing over nearest neighboring spins, and J measures the coupling strength between the neighboring spins. J is divided by 2 to account for double counting of neighboring bonds. For simplicity $J/2$ will be set to unity (in the source program accompanying this chapter) and it is reminded that bonds are counted only once in calculating the sum. We consider regular cubic lattices where one spin is surrounded by 6 nearest neighbors. Equation (4.4) is the objective function to be minimized by simulated annealing method. With periodic boundary conditions, it is seen that this 3-dimensional lattice has $3 \times N^3$ nearest bonds, where N is the number of spins per dimension. The ground state energy (minimal energy) of the system is therefore $-3 \times N^3$.

Note that minimizing can simply be turned into maximizing by multiplying the objective function by -1, or by taking a reciprocal of it, depending on which makes more sense. Problems of maximizing is therefore equivalent to those of minimizing.

4.4 Cooling Schedule

From the behavior of the magnetization of a ferromagnet in Eq. (4.3), we notice that the system is going through a *system-wise* transition near the critical temperature. When we try to find the minimum function value by the method of simulated annealing, it helps to define an *order parameter* suitable for the problem at hand. We then monitor the change in the order parameter with temperature and 'slow down' the cooling procedure when this order parameter is experiencing a transition. It then assures that the minimum function value obtained by the annealing is a global one.

Imagine a system in thermal equilibrium at an initial temperature. The relative chance of the system with an energy of E is determined by the Boltzmann factor of Eq. (4.1). When the temperature is lowered (by a finite amount), the system starts relaxing and, after some 'enough' iterations, new thermal equilibrium is warranted by the Metropolis algorithm. When the system is in thermal equilibrium, order parameters, such as the spin-spin correlation length, and other interesting quantifies can be calculated by averaging over the configurations at the particular temperature. After some experimentation, we can hopefully find an adequate order parameter and its critical temperature.

4.5 3-Dimensional Plot and Animation

In passing, we show in this section how to map 3-dimensional geometrical objects onto 2-dimensional displays.

Imagine a cube is located between our eyes and the screen. When the cube is very far away from the eyes, and thus close to the screen, the angles of the cube corners subtended at our eyes are nearly the same and the subtending edges appear of equal length. As the cube is drawn closer to the eyes, the subtended angles differentiate and the edges closer to the eyes look longer on the screen. The perspective projection is illustrated in Figure 4.1. The task for a 3-dimensional plot is therefore to calculate the subtended angles and then extrapolate the lines of sight up to the screen. This is demonstrated in

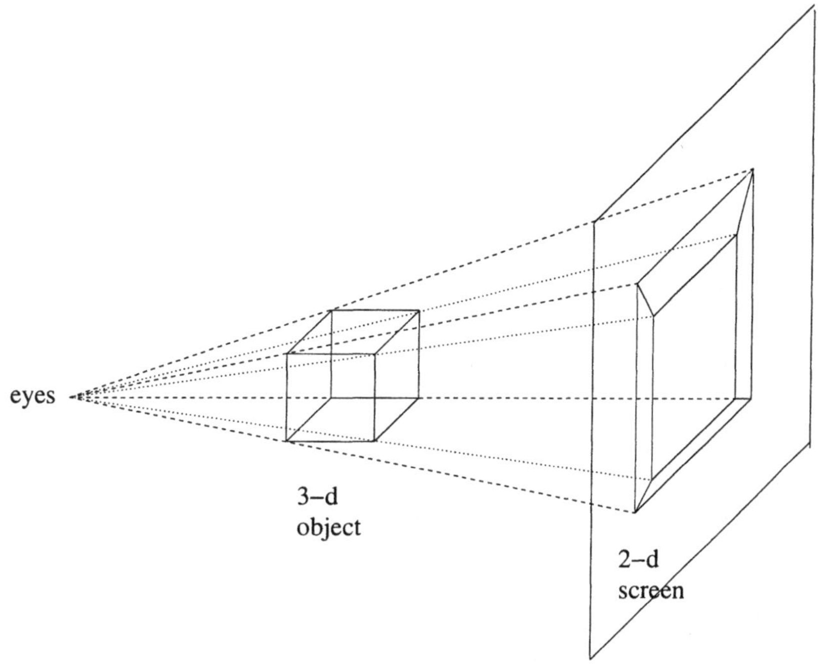

Figure 4.1. Geometry of perspective projection

class `Renderer`, which instantiates the `Matrix` class (of Chapter 1) whose `rotation()` method carries out coordinate transformations (rotations of the cube) along any axis.

The `Observable` class and `Observer` interface, provided in Java's utility package, are found handy in animation. Any class, which `extends Observabl e`, notifies the observer(s) whenever its state is changed. If, for example, the programmer's plotting class implements the `Observer` interface, continuous changes of the coordinates are notified and thus graphed in the canvas by the plotting class. Classes `Animate` and `Renderer` in this chapter show such an implementation.

4.6 An Annealing Example

We are seeking the solution which minimizes the energy of the 3 dimensional Ising model, Eq. (4.4), by the method of simulated annealing. Class `Spin` of Listing 4.1, which contains the `main()` method, sets up the frame and menu bar for the application. The method of simulated annealing is implemented in `Anneal.java` of Listing 4.2. To start minimizing, select `control` from the menu bar. There pops up a dialog box (Figure 4.2) through which the user can alter the annealing parameters, including the number of temperature lowering steps, number of iterations per temperature, starting and ending temperature. The number of spins per dimension can also be changed. This interactive feature helps tame the cooling process. For example, for $N = 8$, a starting temperature of 2.0, a stopping temperature of 1.0, 3 steps of temperature cooling, and a 5,000 iteration per temperature step seem adequate. For a different N, a different set of parameters may find to do the job just as well.

```
/*      Sun-Chong Wang
        TRIUMF
        4004 Wesbrook Mall
        Vancouver, V6T 2A3
        Canada
        e-mail: wangsc@triumf.ca

        Spin.java sets up the frame and menu for the
        simulated annealing application */

import java.awt.*;
import java.lang.*;
import java.awt.event.*;

public class Spin extends Frame implements ActionListener {
 Animate animate;
 SADialog sadlg;
 VDialog vdlg;
 Renderer rendering;
 Panel mypanel;

 public static void main(String args[]) {
        Spin demo = new Spin();
        demo.show();
 }      // end of main

 public Spin() {
        super();
        setTitle("Spin Glass");

        Font font = new Font("Dialog", Font.BOLD, 13);
        setFont(font);
        animate = new Animate(this);
        sadlg = new SADialog(this,"Annealing Control");
        vdlg = new VDialog(this,"Viewing Control");
        rendering = new Renderer(this);
        rendering.addObserver(animate);

        // enables quitting the application by clicking the
        // x button on the upper right corner of the window
        addWindowListener(new WindowAdapter() {
           public void windowClosing(WindowEvent e)
              {System.exit(0);} });

        addMenus();
        mypanel = new Panel();
        mypanel.setLayout(new BorderLayout());
        mypanel.add(animate, BorderLayout.CENTER);
        add(mypanel);

        pack();
        setSize(new Dimension(500,500));
 }      // end of constructor

 private void addMenus() {
        MenuBar mymenubar = new MenuBar();

        Menu myfile = new Menu("File");
```

```
        myfile.add("Quit");
        Menu view = new Menu("View");
        view.add("Show");
        view.add("Hide");
        Menu anneal = new Menu("Anneal");
        anneal.add("Control");

        myfile.addActionListener(this);
        view.addActionListener(this);
        anneal.addActionListener(this);

        mymenubar.add(myfile);
        mymenubar.add(view);
        mymenubar.add(anneal);
        setMenuBar(mymenubar);
 }      // end of addMenus

 // action handler
 public void actionPerformed(ActionEvent e) {
        String action_is = e.getActionCommand();
        if (action_is.equals("Quit")) {
           System.exit(0);
        } else if (action_is.equals("Show")) {
           vdlg.show();
           vdlg.toFront();
        } else if (action_is.equals("Hide")) {
           vdlg.hide();
        } else if (action_is.equals("Control")) {
           sadlg.show();
           sadlg.toFront();
        }
 }      // end of actionPorformed
}       // end of Spin class
```

Listing 4.1 Spin.java

```
/*      Sun-Chong Wang
        TRIUMF
        4004 Wesbrook Mall
        Vancouver, V6T 2A3
        Canada
        e-mail: wangsc@triumf.ca

        Anneal.java embodies the method of stimulated annealing */

import java.lang.*;
import java.util.*;
import java.util.Random;

public class Anneal {
 Random rand;          // random number generator to be discussed in Ch. 7
 int nTemperatures;    // number of temperatures
 int niters;           // number of iterations at each temperature
 int move;             // number of Metropolis moves
 double fquit;         // quit if function reduced to this amount
 double starttemp;     // starting temperature
 double stoptemp;      // stopping temperature
 int N,nvars;          // number of total spins
 double[] center, spin, bestspin;
 double bestfval;
 double[][][] sJ;      // to store coupling constants
 double T;

 SADialog parent;
```

```
public Anneal(SADialog parent) {
        this.parent   = parent;
        rand          = new Random();
        T             = 1.0;
        nTemperatures = 10;
        niters        = 1000;
        fquit         = -Double.MAX_VALUE;
        N             = 0;
        System.out.println("random number = "+rand.nextDouble());
}       // end of Anneal class constructor

public void initialize(int N) {
        if (N != this.N) {
           this.N   = N;
           nvars    = N*N*N;     // 3-d Ising model
           sJ       = new double[2*N+1][2*N+1][2*N+1];
           spin     = new double[nvars];
           center   = new double[nvars];
           bestspin= new double[nvars];
           initJ();              // initialize the coupling constants
        }
        for (int i=0; i<nvars; i++) { // initialize trial spins
            if (rand.nextDouble() < 0.5) spin[i] = -1.0;
            else spin[i] = 1.0;
            bestspin[i] = spin[i];
        }
        bestfval = SpinGlass();
        System.out.println("start from E = "+SpinGlass());
}

// calculates the Hamiltonian of the Ising system, i.e. Eq. (4.4)
public double SpinGlass() {
        int i,j,k,N2;
        double E;

        N2 = 2*N;
        E = 0.0;

        // translating from spin[] to sJ[][][]
        k = 0;
        while (k < N) {
              j = 0;
              while (j < N) {
                    i = 0;
                    while (i < N) {
                          sJ[2*i][2*j][2*k] = spin[i+((j+(k*N))*N)];
                          i += 1;
                    }
                    j += 1;
              }
              k += 1;
        }
        // periodic boundary conditions
        k = 0;
        while (k < N2) {
              j = 0;
              while (j < N2) {
                    sJ[N2][j][k] = sJ[0][j][k];
                    j += 2;
              }
              k += 2;
        }
        k = 0;
        while (k < N2) {
              i = 0;
              while (i < N2) {
                    sJ[i][N2][k] = sJ[i][0][k];
```

```
                    i += 2;
            }
            k += 2;
    }
    j = 0;
    while (j < N2) {
            i = 0;
            while (i < N2) {
                    sJ[i][j][N2] = sJ[i][j][0];
                    i += 2;
            }
            j += 2;
    }
    // Eq. (4.4)
    k = 0;
    while (k < N2) {
            j = 0;
            while (j < N2) {
                    i = 0;
                    while (i < N2) {
                            E -=
                            sJ[i][j][k]*(sJ[i+1][j][k]*sJ[i+2][j][k]
                                      + sJ[i][j+1][k]*sJ[i][j+2][k]
                                      + sJ[i][j][k+1]*sJ[i][j][k+2]);
                            i += 2;
                    }
                    j += 2;
            }
            k += 2;
    }
    return E;
}   // end of function SpinGlass

public double[] getSpins() {
        return bestspin;
}

public void go() {             // the annealing process
        int i,j,k;
        double fval, oldfval, df;
        double prob, ratio, factor;

        for (i=0; i<nvars; i++) center[i] = spin[i];

        oldfval = SpinGlass();
        bestfval = oldfval;

        T = starttemp;
        ratio = stoptemp/starttemp;
        if (nTemperatures != 1)
           factor = Math.exp(Math.log(ratio)/(nTemperatures-1));
        else factor = 1.0;

        for (i=0; i<nTemperatures; i++) { // temp reduction loop
            move = 0;
            for (j=0; j<niters; j++) {     // iterations per temp loop
                perturb();
                fval = SpinGlass();
                df = fval - oldfval;
                if (fval < bestfval) {
                   bestfval = fval;
                   for (k=0; k<nvars; k++) bestspin[k] = spin[k];
                }

                if (df <= 0.0) {            // if improved
                   for (k=0; k<nvars; k++) center[k] = spin[k];
                   oldfval = fval;
                } else {                    // else Metropolis move
```

```
                    prob = Math.exp(-df/T);
                    if (rand.nextDouble() < prob) {
                       for (k=0; k<nvars; k++) center[k] = spin[k];
                       move += 1;
                       oldfval = fval;
                    }
                 }

                 if (bestfval <= fquit) break;
              }  // end of j loop
              // plot the spins after every temperature
              parent.parent.rendering.go(getSpins(),N);
              parent.parent.rendering.notifyObservers(
                             parent.parent.animate);
              if (bestfval <= fquit) break;

              System.out.println("there are " + move +
              " Metropolis moves at T = "+T+" and E = "+bestfval);
              // monitoring the number of Metropolis moves help
              // determine the appropriate range of temperature
              // (start and stop temperature) for the Boltzmann's
              // factor
              T *= factor;
           }  // end of i loop

 }       // end of method go

 public void initJ() { // Ising model
         int i,j,k,N2;
         Random ran = new Random();

         ran.setSeed(1L);
         N2 = 2*N;

         // initializing couplings
         k = 0;
         while (k < N2) {
               j = 0;
               while (j < N2) {
                     i = 1;
                     while (i < N2) {
                           sJ[i][j][k] = 1.0;
/* use the following couplings will turn an Ising into spin glass
                           if (ran.nextDouble()<1.0) sJ[i][j][k]=-1.0;
                           else sJ[i][j][k] = 1.0;
*/
                           i += 2;
                     }
                     j += 2;
               }
               k += 2;
         }
         k = 0;
         while (k < N2) {
               i = 0;
               while (i < N2) {
                     j = 1;
                     while (j < N2) {
                           sJ[i][j][k] = 1.0;
// spin glass                if (ran.nextDouble()<1.0) sJ[i][j][k]=-1.0;
//                           else sJ[i][j][k] = 1.0;
                           j += 2;
                     }
                     i += 2;
               }
               k += 2;
```

```
        }
        j = 0;
        while (j < N2) {
                i = 0;
                while (i < N2) {
                        k = 1;
                        while (k < N2) {
                                sJ[i][j][k] = 1.0;
// spin glass                     if (ran.nextDouble()<1.0) sJ[i][j][k]=-1.0;
//                                else sJ[i][j][k] = 1.0;
                                k += 2;
                        }
                        i += 2;
                }
                j += 2;
        }
 }      // end of method initJ

 public void perturb() {
        int i,j;

        j = rand.nextInt(nvars);
        for (i=0; i<j; i++) spin[i] = center[i];
        spin[j] = -center[j];
        for (i=j+1; i<nvars; i++) spin[i] = center[i];
 }      // end of method perturb
}       // end of Anneal class
```

Listing 4.2 Anneal.java

Method `SpinGlass()` in `Anneal` calculates the energy of the system defined in Eq. (4.4). The spins are stored in a one dimensional array, `spin[]`, and manipulated in the annealing procedure. It is translated back to a 3-dimensional array, `sJ[][][]`, when the energy is calculated. `sJ[][][]` also holds the coupling constants between neighboring spins and if their signs are randomly assigned, the Ising system is turned into a spin glass system which is known to contain plenty of local minima. The one dimensional array storing scheme makes the implementation independent of the dimension of the system. The Ising problem can also be solved by the genetic algorithm of Chapter 6 where possible solutions are represented by chromosomes, which are envisioned as one dimensional arrays.

A special note is devoted to method `perturb()` where shuffling is done. There are N^3 independent lattice spins each of which can point up or down. We start from a configuration of random spin orientations. The starting energy of the system is therefore close to zero. To try a new configuration, a single spin is randomly selected from the N^3 spins and its orientation is inverted. If the trial configuration yields a lower energy, it's adopted and becomes the configuration which spawns the next trial configuration. On the other hand, if it gives a higher energy, whether or not it is accepted is determined by the Metropolis algorithm. Shuffling operation is critical to the performance of the algorithm. Different problems, such as the traveling salesman problem, are owed clever ways of shuffling for efficient optimization. We show an updating scheme for fast error function minimization in Section 7.

Listing 4.3 is the class which pops up a separate window (dialog box) for user input and program output (Figure 4.2). With the interface, the user can experiment her way toward efficient cooling.

```
/*      Sun-Chong Wang
```

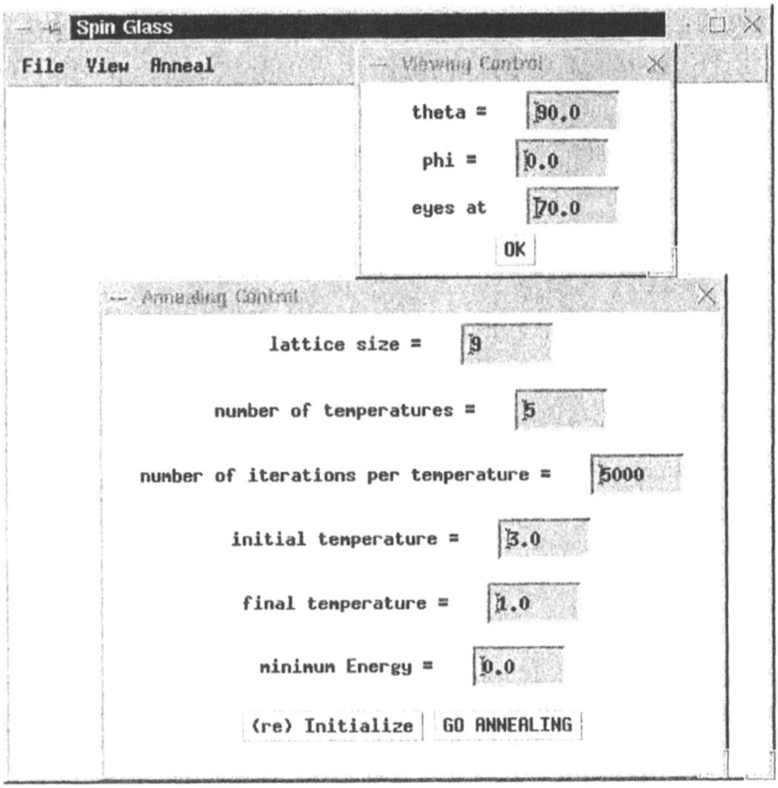

Figure 4.2. Dialog boxes for interaction

```
        TRIUMF
        4004 Wesbrook Mall
        Vancouver, B.C. V6T 2A3
        Canada
        e-mail: wangsc@triumf.ca

        SADialog defines a dialogue box to hold parameters
        for the simulated annealing minimization */

import java.lang.*;
import java.awt.*;
import java.awt.event.*;

class SADialog extends Dialog implements ActionListener {

 TextField NTF, nTTF, nITF, TiTF, TfTF, ETF;
 Integer NI, nTI, nII;
 Double TiD, TfD, ED;
 double Ti, Tf, energy;
 int N, nTemperatures, nIterations;

 Anneal annealing;
 Spin parent;

 public SADialog(Spin parent, String text) {
```

```
super(parent,text,false);
setBackground(Color.white);
this.parent = parent;

annealing = new Anneal(this);

// set default values
N = 9;                     // 9 spins per dimension
nTemperatures = 5;         // 5 temperatures
nIterations   = 5000;      // 5000 iterations per temp
Ti            = 3.0;       // initial temp
Tf            = 1.0;       // stopping temp
energy        = 0.0;

NI = new Integer(N);
nTI = new Integer(nTemperatures);
nII = new Integer(nIterations);
TiD = new Double(Ti);
TfD = new Double(Tf);
ED  = new Double(energy);

setLayout(new GridLayout(7,1));  // need a grid of 7

// define the 7 panels
Panel Npanel = new Panel();
Panel nTpanel = new Panel();
Panel nIpanel = new Panel();
Panel Tipanel = new Panel();
Panel Tfpanel = new Panel();
Panel Epanel  = new Panel();
// one more later

NTF = new TextField(Integer.toString(N),6);
Npanel.add(new Label("lattice size = ",Label.LEFT));
Npanel.add(NTF);
add(Npanel);

nTTF = new TextField(Integer.toString(nTemperatures),6);
nTpanel.add(new Label("number of temperatures = ",
                                 Label.LEFT));
nTpanel.add(nTTF);
add(nTpanel);

nITF = new TextField(Integer.toString(nIterations),6);
nIpanel.add(new Label("number of iterations per
                       temperature = ", Label.LEFT));
nIpanel.add(nITF);
add(nIpanel);

TiTF = new TextField(Double.toString(Ti),6);
Tipanel.add(new Label("initial temperature = ",Label.LEFT));
Tipanel.add(TiTF);
add(Tipanel);

TfTF = new TextField(Double.toString(Tf),6);
Tfpanel.add(new Label("final temperature = ",Label.LEFT));
Tfpanel.add(TfTF);
add(Tfpanel);

ETF = new TextField(Double.toString(energy),6);
Epanel.add(new Label("minimum Energy = ",Label.LEFT));
Epanel.add(ETF);
add(Epanel);
```

```
        Panel bb = new Panel();
        Button button1 = new Button("(re) Initialize");
        Button button2 = new Button("GO ANNEALING");
        button1.addActionListener(this);
        button2.addActionListener(this);
        bb.add(button1);
        bb.add(button2);

        add(bb);

        pack();
        setSize(new Dimension(400,300));
 }      // end of SADialog constructor

 // action handler
 public void actionPerformed( ActionEvent e) {
        if ("(re) Initialize".equals(e.getActionCommand())) {
           N = NI.parseInt(NTF.getText());
           nTemperatures = nTI.parseInt(nTTF.getText());
           nIterations = nII.parseInt(nITF.getText());
           Ti = TiD.parseDouble(TiTF.getText());
           Tf = TfD.parseDouble(TfTF.getText());

           annealing.nTemperatures = nTemperatures;
           annealing.niters = nIterations;
           annealing.starttemp = Ti;
           annealing.stoptemp = Tf;
           annealing.initialize(N);
           energy = annealing.bestfval;
           ETF.setText(ED.toString(energy));

           parent.rendering.go(annealing.getSpins(),N);
           parent.rendering.notifyObservers(parent.animate);
        }
        if ("GO ANNEALING".equals(e.getActionCommand())) {
           N = NI.parseInt(NTF.getText());
           nTemperatures = nTI.parseInt(nTTF.getText());
           nIterations = nII.parseInt(nITF.getText());
           Ti = TiD.parseDouble(TiTF.getText());
           Tf = TfD.parseDouble(TfTF.getText());

           annealing.nTemperatures = nTemperatures;
           annealing.niters = nIterations;
           annealing.starttemp = Ti;
           annealing.stoptemp = Tf;

           Message runningBox = new Message(parent,
                                    "Annealing","Running");
           runningBox.show();
           // the above small popup message box persists as long as
           // the annealing is still running
           annealing.go();
           energy = annealing.bestfval;
           ETF.setText(ED.toString(energy));
           runningBox.dispose();
        }
 }      // end of actionPerformed
}       // end of SADialog Class
```

Listing 4.3 SADialog.java

Figure 4.3 shows an example of the starting configuration of the Ising model. Blue and red line segments represent spins of opposite direction. Figures 4.4 to 4.6 are screen shots of the configurations during annealing. Figures 4.7 and 4.8 are the two possible minimum energy configurations: all blue and all

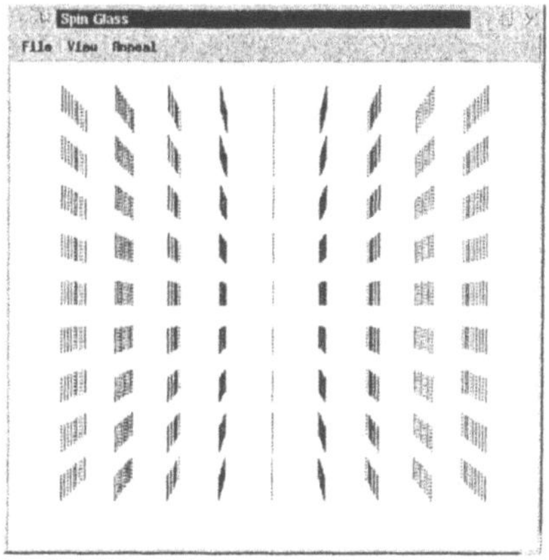

Figure 4.3. Initial configuration

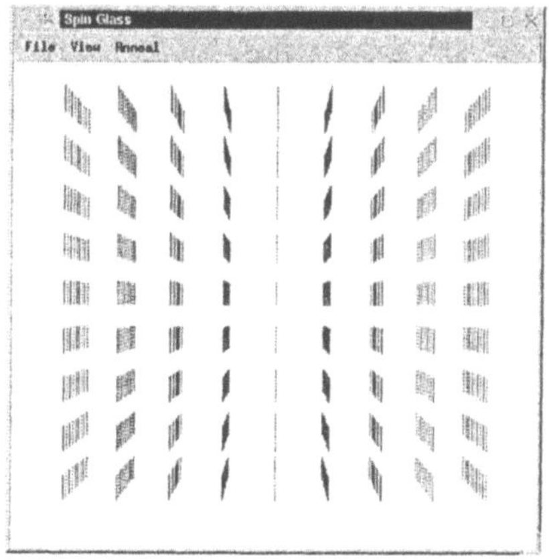

Figure 4.4. Cooling down

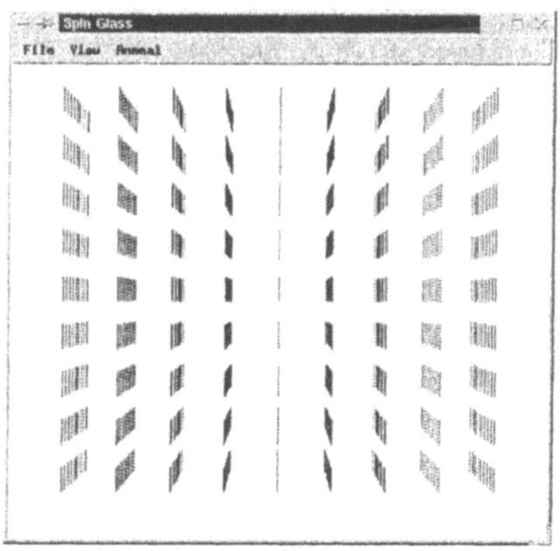

Figure 4.5. Cooling down continued

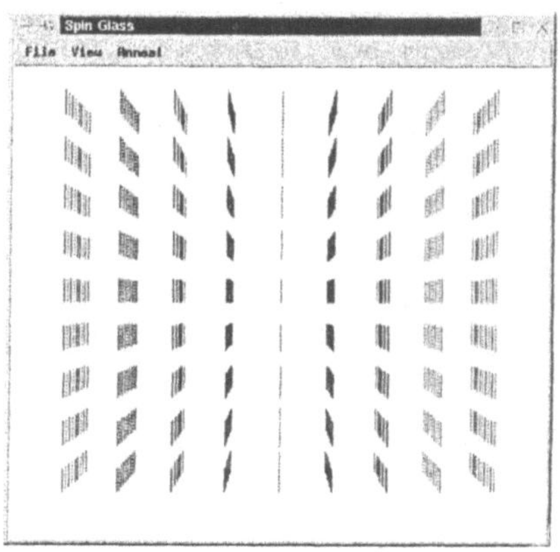

Figure 4.6. Cooling down continued

red. Figures 4.9 to 4.12 shows some interesting domain forming configurations during simulation.

Listing 4.4 shows the class for rendering 3-dimensional scenes onto a 2-dimensional display. Similarly the user can change the viewing angle and distance through a pop up dialog. Different renderings are shown in Figures 4.13 to 4.16 and the screen shot of the dialog boxes are shown in Figure 4.2. The program also outputs numerical numbers on the shell terminal where the application is launched:

```
there are 473 Metropolis moves at T = 3.0 and E = -1139.0
there are 191 Metropolis moves at T = 2.2795070569547775 and E = -1555.0
there are 74 Metropolis moves at T = 1.7320508075688772 and E = -1903.0
there are 9 Metropolis moves at T = 1.3160740129524924 and E = -2151.0
there are 0 Metropolis moves at T = 0.9999999999999999 and E = -2187.0
```

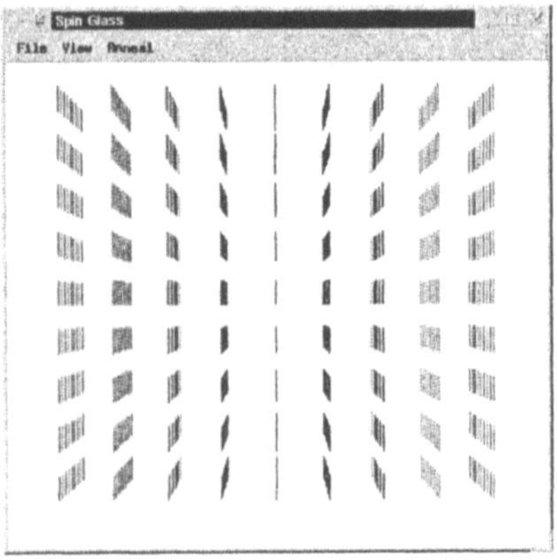

Figure 4.7. Ground state (all blue)

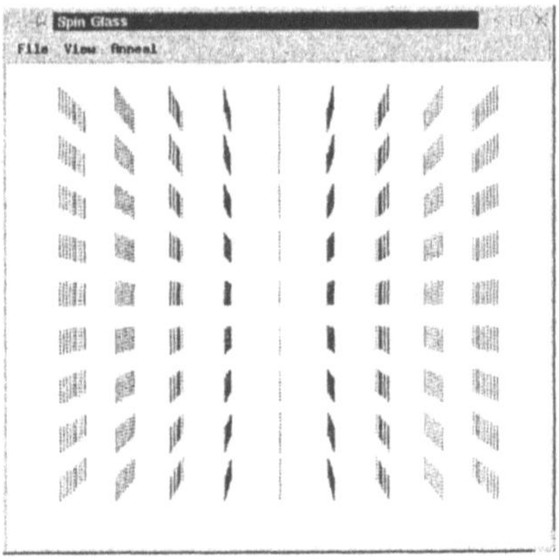

Figure 4.8. The other ground state

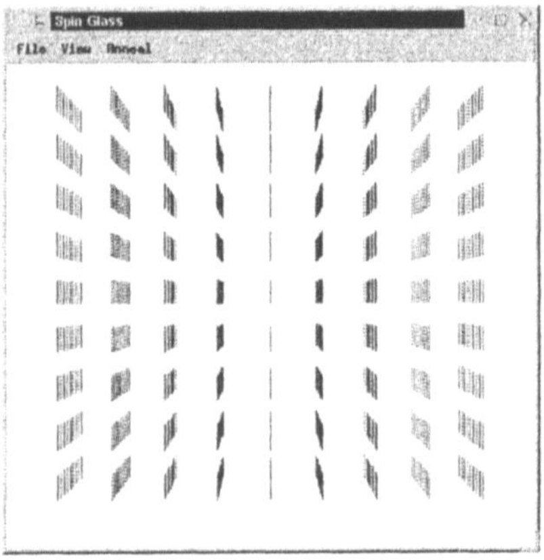

Figure 4.9. An example of state at low temperature

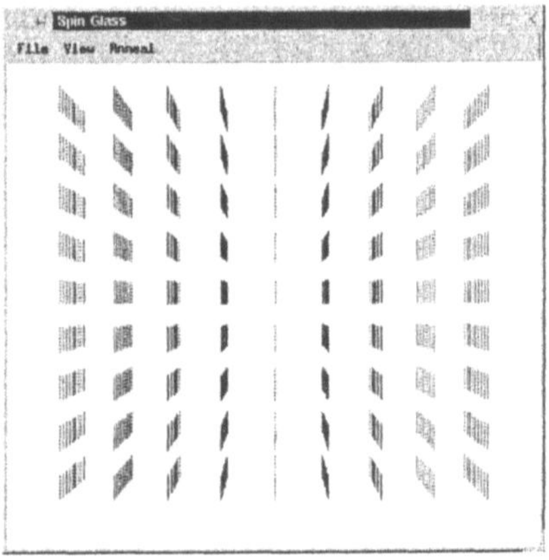

Figure 4.10. An example of state at low temperature

```
/*      Sun-Chong Wang
        TRIUMF
        4004 Wesbrook Mall
        Vancouver, V6T 2A3
        Canada
        e-mail: wangsc@triumf.ca

        Renderer.java rotates and projects a 3-d object */

import java.util.*;
import java.lang.*;

class Renderer extends Observable {

 double eye_x, eye_y, eye_z, screen_z;
 double[][][] site;
```

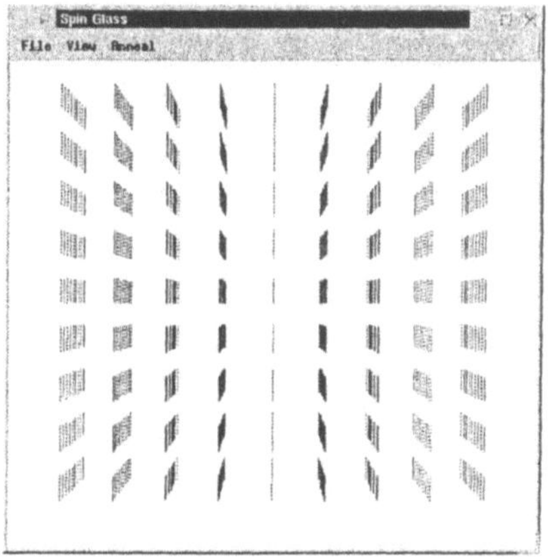

Figure 4.11. An example of state at low temperature

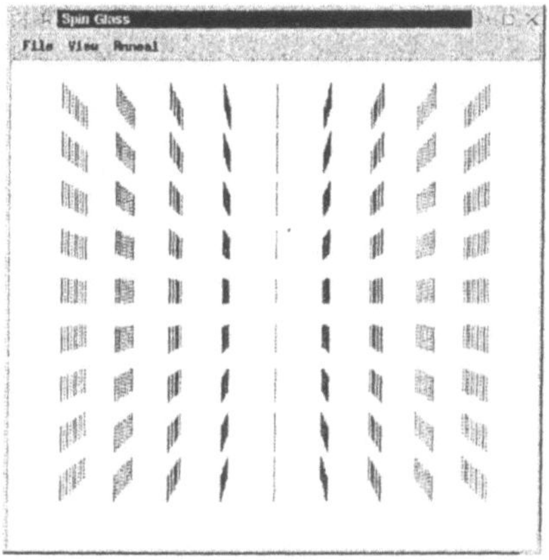

Figure 4.12. An example of state at low temperature

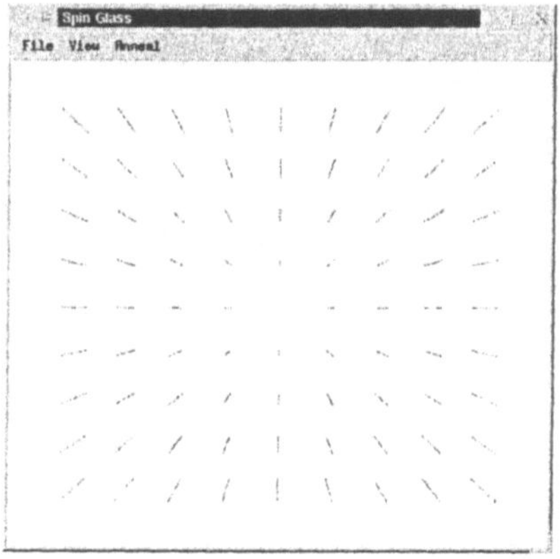

Figure 4.13. Different viewing angle

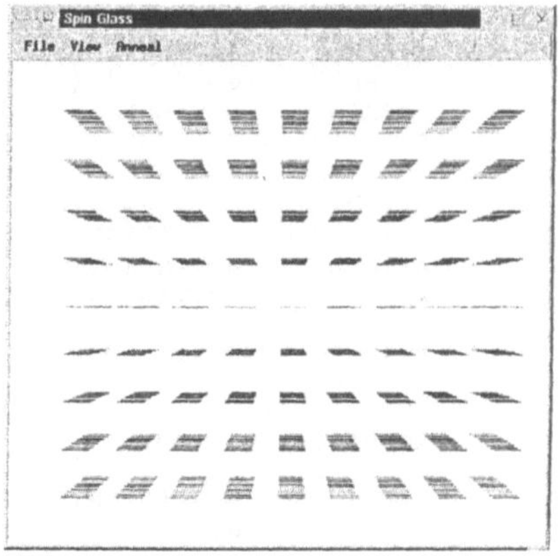

Figure 4.14. Different viewing angle

```
Matrix orient, orient1, orient2, Identity;
double theta, phi;

Spin parent;

public Renderer(Spin parent) {
        super();
        this.parent = parent;

        eye_x = 0.0;              // viewer's location
        eye_y = 0.0;
        eye_z = 70.0;
        screen_z = -100.0;        // screen's location
        theta = 90.0;
        phi = 0.0;
```

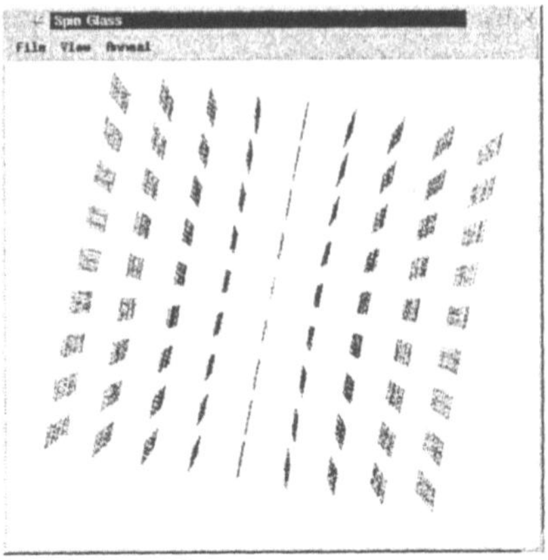

Figure 4.15. Different viewing distance

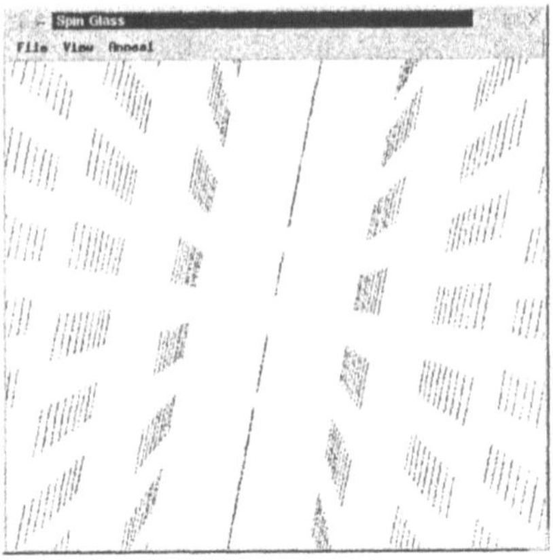

Figure 4.16. Different viewing distance

```
        orient = new Matrix(3,3);
        orient1 = new Matrix(3,3);
        orient2 = new Matrix(3,3);
        Identity = new Matrix(3,3);
        for (int i=0; i<3; i++) Identity.M[i][i] = 1.0;

        orient1.M = Identity.rotation(2, Math.toRadians(phi));
        orient2.M = Identity.rotation(0, Math.toRadians(theta));
        orient.M = orient1.times(orient2.M);
}       // end of constructor

public void go(double[] spin, int N) {
        double slopex, slopey;
        int i,j,k,l;
        double[][][] tmp;

        site = new double[spin.length][2][3];
        tmp  = new double[spin.length][2][3];

        // loading data (3-d coordinates of the spins)
        for (k=0; k<N; k++) {
           for (j=0; j<N; j++) {
                for (i=0; i<N; i++) {
                    l = i + j*N + k*N*N;
                    site[l][0][0] = i - N/2;  // x
                    site[l][0][1] = j - N/2;  // y
                    site[l][0][2] = k - N/2;  // z
                    site[l][1][0] = site[l][0][0];
                    site[l][1][1] = site[l][0][1];
                    // length of the spin is 0.5
                    site[l][1][2] = site[l][0][2] + 0.5;
                }
           }
        }
        // rotation matrix
        orient1.M = Identity.rotation(2, Math.toRadians(phi));
        orient2.M = Identity.rotation(0, Math.toRadians(theta));
        orient.M = orient1.times(orient2.M);

        // rotating
        for (l=0; l<2; l++) {
```

```
            for (i=0; i<spin.length; i++) {
                for (j=0; j<3; j++) {
                    tmp[i][1][j] = 0.0;
                    for (k=0; k<3; k++) {
                        tmp[i][1][j] += orient.M[j][k]*site[i][1][k];
                    }
                }
            }
        }

        // projecting
        for (j=0; j<2; j++) {
            for (i=0; i<spin.length; i++) {
                // find the angles first
                slopex = (tmp[i][j][0]-eye_x)/(tmp[i][j][2]-eye_z);
                slopey = (tmp[i][j][1]-eye_y)/(tmp[i][j][2]-eye_z);
                // then extrapolate to the screen
                site[i][j][0] = slopex*(screen_z-eye_z);
                site[i][j][1] = slopey*(screen_z-eye_z);
            }
        }
        parent.animate.spin = spin;
        parent.animate.site = site;
        setChanged();
    }   // end of method go
}       // end of class Renderer
```

Listing 4.4 Renderer.java

4.7 Minimization of Functions of Continuous Variables

Simulated annealing can also be used to minimize functions whose parameters are continuously varying. It is of particular use when the function is nonlinear in the parameters. A commonest rite for a researcher is to fit data to a model. Suppose the model, g, can be written as a sum of terms like,

$$g(a, b, c, d, e) = g_1(a, b) + g_2(b, c, d) + g_3(c, d, e), \tag{4.5}$$

where the model parameters, a, b, c, d, e, are real-valued. g_i can be thought of as the function to model expression (activity) levels of gene i in molecular biology. Since a gene can influence other genes, and vice versa, the parameters, measuring the strength of interaction, are highly coupled among the functions.

To estimate the best parameter values, we, by tuning values of a, b, c, d, e, minimize the sum of squared errors, χ^2,

$$\begin{aligned} \chi^2(a, b, c, d, e) &\equiv \left(g_1(a, b) - g_1'\right)^2 + \left(g_2(b, c, d) - g_2'\right)^2 \\ &\quad + \left(g_3(c, d, e) - g_3'\right)^2 \\ &\equiv \chi^2_{g_1}(a, b) + \chi^2_{g_2}(b, c, d) + \chi^2_{g_3}(c, d, e), \end{aligned} \tag{4.6}$$

where g_1', g_2', g_3' are measured values of g_1, g_2, g_3 (by DNA microarray in the example of gene expression). Here, the χ^2 function is the objective function of the simulated annealing. The annealing again starts from a set of initial trial values of a,b, c,d,e. We then randomly select one parameter out of the five, say b, and change its value by adding a perturbing term $\xi\epsilon$,

$$b \rightarrow b + \xi\epsilon, \tag{4.7}$$

where ξ is a small number whose magnitude is characteristic of the problem at hand. ϵ is a draw from either a random uniform distribution between -1 and 1 or a random Gaussian distribution with zero mean and unit standard deviation. We will detail random number generation in Chapter 7. The size of ξ can be made decreasing with the temperature. Its initial magnitude and the value of the starting temperature can be chosen so that the probability of unfavorable transition of Eq. (4.2) is about few percent. The rest proceeds with the standard annealing procedure.

When the number of parameters gets huge, minimization becomes slow. To speed up the search, we can update a group of parameters at once. The question arises as to how to group parameters. Decomposition in Eq. (4.6) offers a hint: (a, b), (b, c, d), and (c, d, e) are the appropriate choice. Moreover, since individual error contributions, $\chi^2_{g_1}, \chi^2_{g_2}, \chi^2_{g_3}$ (or $\chi^2_{g_1}/g_1'^2, \chi^2_{g_2}/g_2'^2, \chi^2_{g_3}/g_3'^2$), are available, we, in the perturbing step, update only the group of parameters that contribute *most* to the total χ^2. The algorithm thus works on the parameters that are still far off the optimal values while keeping intact the parameters that have already come close to their optimal values.

4.8 Summary

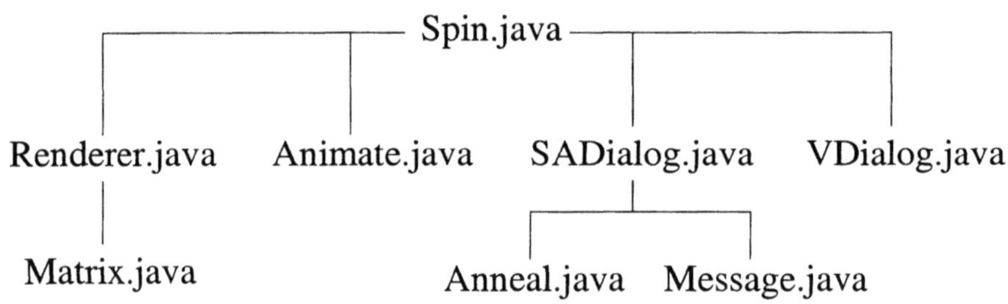

Figure 4.17. Source programs of the simulated annealing application

`Animate.java` and `Message.java` can be obtained by modifying the templates in the appendix. `VDialog.java` is a simplified version of `SADialog.java`. `Matrix.java` is the one in Chapter 1.

We found the variable values (spins) at which the objective function (energy of a 3-dimensional Ising model) is minimum by the method of simulated annealing. The cooling sequence of the method for an efficient search is often dependent on the problem. We provided a dialog box which makes the engineering process easy. We showed how 3-dimensional scenes are projected on 2-dimensional screens and provided a class for rotating and rendering the 3-dimensional object. Simulated annealing is also applicable to functions of continuous variables.

Simulated annealing is a powerful method for optimization. It is built in a mechanism which prevents the search from being trapped in suboptimal regions of the configuration space. The method imitates how physical systems in nature work in settling down to their stable states.

4.9 References and Further Reading

The 1-dimensional Ising model was solved in, E. Ising, "Beitrag zur theorie des ferromagnetismus", *Zeits. fur Phys.*, **31** (1925) 253-258

Theories of equilibrium thermodynamics can be found in the textbook, K, Huang, "Statistical Mechanics", John Wiley & Sons, New York (1987)

Simulated annealing and thermodynamics were discussed in, S. Kirkpatrick, C.D. Gelatt, and M.P. Vecchi, "Optimization by Simulated Annealing", *Science*, **220** (1983) 671-680

The Metropolis algorithm appeared in, N. Metropolis, A.W. Rosenbluth, M.N. Rosenbluth, E.H. Teller, and E. Teller, "Equation of State Calculations by Fast Computing Machines", *Journal of Chemical Physics*, **21** (1953) 1087-1091

Chapter 5

ARTIFICIAL NEURAL NETWORK

Inspired by the sophisticated functionality of human brains where hundreds of billions of interconnected neurons process information in parallel, researchers have successfully tried demonstrating certain levels of intelligence on silicon. Examples include language translation and pattern recognition software. While simulation of human consciousness and emotion is still in the realm of science fiction, we, in this chapter, consider artificial neural networks as universal function approximators. Especially, we introduce neural networks which are suited for time series forecasts.

5.1 Introduction

An artificial neural network (or simply neural network) consists of an input layer of neurons (or nodes, units), one or two (or even three) hidden layers of neurons, and a final layer of output neurons. Figure 5.1 shows a typical architecture, where lines connecting neurons are also shown. Each connection is associated with a numeric number called *weight*. The output, h_i, of neuron i in the hidden layer is,

$$h_i = \sigma\Big(\sum_{j=1}^{N} V_{ij}x_j + T_i^{hid}\Big), \tag{5.1}$$

where $\sigma()$ is called activation (or transfer) function, N the number of input neurons, V_{ij} the weights, x_j inputs to the input neurons, and T_i^{hid} the threshold terms of the hidden neurons. The purpose of the activation function is, besides introducing nonlinearity into the neural network, to bound the value of the neuron so that the neural network is not paralyzed by divergent neurons. A common example of the activation function is the sigmoid (or logistic) function

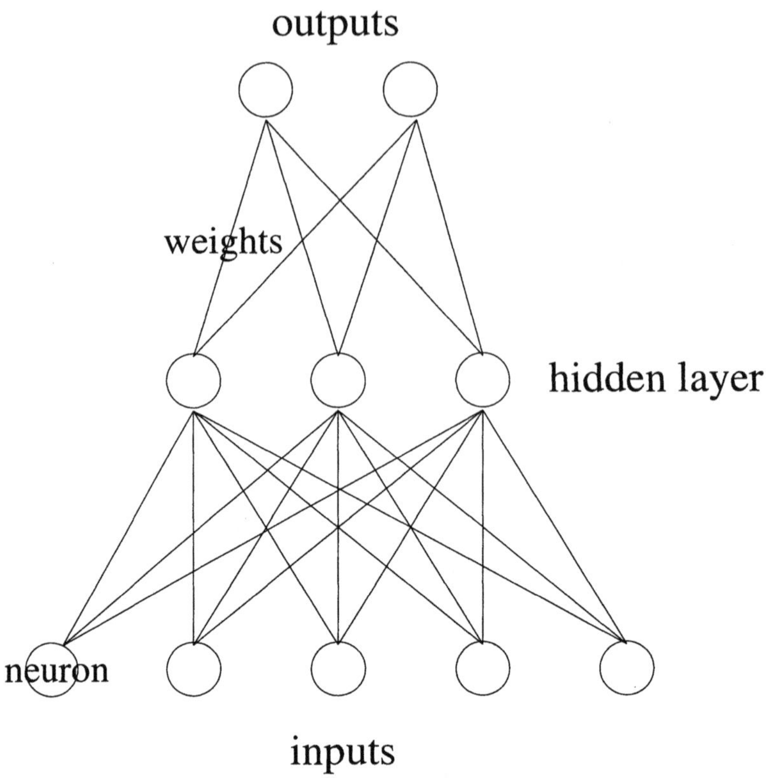

Figure 5.1. Architecture of a neural network

defined as (Figure 5.2),

$$\sigma(u) = \frac{1}{1 + \exp(-u)}. \tag{5.2}$$

Other possible activation functions are arc tangent and hyperbolic tangent. They have similar response to the inputs as the sigmoid function, but differ in the output ranges.

It has been shown that a neural network constructed the way above can approximate any computable function to an arbitrary precision. Numbers given to the input neurons are independent variables and those returned from the output neurons are dependent variables to the function being approximated by the neural network. Inputs to and outputs from a neural network can be binary (such as yes or no) or even symbols (green, red, ...) when data are appropriately encoded. This feature confers a wide range of applicability to neural networks.

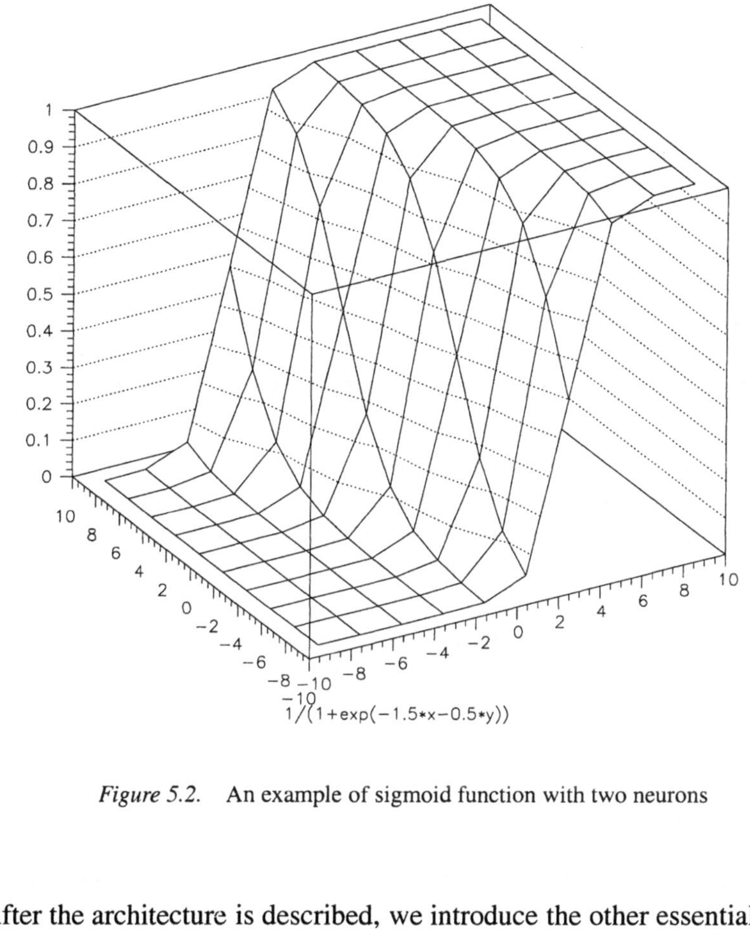

Figure 5.2. An example of sigmoid function with two neurons

After the architecture is described, we introduce the other essential ingredient of a neural network application, namely, *training*. Similar to human learning by examples, a neural network is trained by presenting to it a set of input data called *training set*. The desired outputs of the training data are known so that the aim of the training is to minimize, by adjusting the weights between the connected neurons, an *error function* which is normally the sum of the squared differences between the neural network outputs and the desired outputs.

If we are experimenting architectures of neural networks, an independent data set called *validation set* can be applied to the trained neural networks. The one which performs best is then picked up as the one of choice. After validation, yet another independent data set called *test set* is used to determine the performance level of the neural network which tells how confident we are when using the neural network. It should be understood that a neural network can never learn what is not present in the training set. The size of the training set has therefore to be large enough for the neural network to memorize the features/trends embedded in the training set. On the other hand, if too much unimportant details are contained in the training set, the neural network might waste its resources (weights) fitting the noise. A judicious selection and/or representation of the data are therefore critical to successful implementations of neural networks.

Note that the definitions of validation and test set are reversed among authors of different fields. We have here followed the definition of B.D. Ripley (1996).

This section serves as a general introduction to neural networks. The following sections describe a step-by-step procedure to design a neural network for time series prediction.

5.2 Structural vs. Temporal Pattern Recognition

Conventional neural networks such as the one in Figure 5.1 have proven to be a promising alternative to traditional techniques for structural pattern recognition. In such applications, attributes of the sample data are presented to the neural network at the same time during training. After successful training, the neural network is supposed to be able to categorize the features buried in the training data set.

By contrast, in temporal pattern recognition, features evolve over time. A neural network for temporal pattern recognition is required to both remember and recognize patterns. This additional requirement poses a challenge to not only the design of the neural network architecture but the training procedure and data representation since all these tasks are inter-related; they can in fact be viewed as different aspects of the same underlying problem.

In the next section, we introduce a neural network architecture which has built-in memories and is therefore most suited for tasks of time series prediction.

5.3 Recurrent Neural Network

The network architecture of Figure 5.1 is called feedforward neural network because the direction of information flow is from the input, through the hidden layers, to the output. It has been indicated that for time series prediction recurrent neural networks are more competent. A recurrent neural network is a type of network that has feedbacks similar to the feedback circuitry in electronics. Figure 5.3 shows a simple yet powerful recurrent neural network introduced by J.L. Elman. Because of the delayed feedback, the network now has 'memory'.

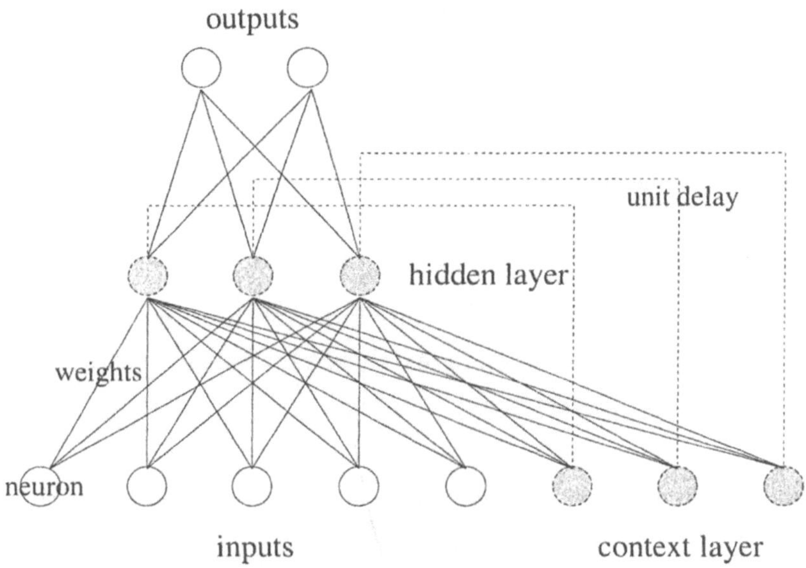

Figure 5.3. Architecture of a recurrent neural network

The output of the neuron i in the hidden layer becomes, instead of Eq. (5.1),

$$h_i(t) = \sigma\Big(\sum_{j=1}^{N} V_{ij}x_j(t) + \sum_{k=1}^{H} W_{ik}h_k(t-1) + T_i^{hid}\Big). \tag{5.3}$$

Time is explicitly indexed above to highlight the delayed feedback. When W_{ik} are all zero, the network reduces to the feedforward neural network of Figure 5.1. The output neurons, $o_i(t)$, at time t are updated as,

$$o_i(t) = \sigma\Big(\sum_{j=1}^{H} Q_{ij}h_j(t) + T_i^{out}\Big), \tag{5.4}$$

where Q_{ij} are the weights between the hidden neurons and the output neurons, and T_i^{out} are the thresholds of the output neurons. For one-unit-time-ahead predictions, we define, in the sense of least squares, the following error function,

$$\chi^2 = \sum_{t=1}^{\tau}\sum_{i=1}^{N}\big(o_i(t) - x_i(t+1)\big)^2, \tag{5.5}$$

where time is discrete and τ is the horizon of the data. For simplicity, we have assumed that the dimensionality of the output vector is the same as that of the input. A properly defined measure of error is one of the few key factors for successful neural network applications. An example will be shown below.

5.4 Steps in Designing a Forecasting Neural Network

We take as an illustration market prediction which is familiar to most readers irrespective of the professional background. The first step in neural network design is the selection of variables. To predict the price of some publicly traded commodity, say Japanese yen, at a future time $t+1$, you would first of all try writing down some equation like,

$$\begin{aligned} \text{yen}(t+1) = \text{yen}(t) + & \\ & f\big(\text{dollar index}(t),\ \text{Euro index}(t),\ \text{Sterling index}(t), \\ & 10 - \text{year bond index}(t),\ \text{Nikkei} - 225(t),\ \text{DJIA}(t), \\ & \text{oil price}(t),\ \cdots\big). \end{aligned} \tag{5.6}$$

Variables of the function $f()$ are the indicators of the market forces at present time t that you believe will affect the price of yen in the currency exchange market. The function $f()$ in Eq. (5.6) is presumably highly nonlinear. And particularly there exit no theories to model the behavior of yen over time. It is this difficulty that leads us to the method of neural networks. The choice of the frequency of the data, namely hourly, daily, or monthly exchange rates, depends on the objective of the researcher. The horizon of the historical data for training is another issue since trends change over time. You have also to make sure that values of the variables have been evaluated in a consistent way over time.

The second step is data representation. Raw data are rarely fed into neural networks without preprocessing/transformation. One of the the common transformations is to take natural logarithm of the change in the variable value. The transformed values form a distribution whose mean and standard deviation can be readily obtained. A scaling operation is then performed so that the range of the data is bounded between 0 and 1 or -1 and 1. A typical scaling is to assign to 1 (0) all values beyond (below) 3 standard deviations away from the mean. Values within 3 standard deviations are then linearly scaled between 0 and 1. Again which of the two ranges to use depends on the activation function. It was however pointed out that sigmoid activation functions were better for neural networks learning average behavior, while hyperbolic tangent worked better if learning involves deviations from the average.

Step 3 is training, validation, and testing. We move forward in time and divide the whole historical data set into training, validation, and test set at a proportion of, say, 7:2:1. The advantage of having the validation set follow the training set is that these data contain the most up-to-date market trends. On the other hand, care has to be taken to make sure that the trained neural network does not favor a subgenre of market trends.

5.5 How Many Hidden Neurons/Layers ?

We have known that recurrent neural networks of Figure 5.3 are suitable for time series prediction. However, to be specific, the next step is to determine the number of hidden layers and the number of neurons in a hidden layer.

There are as many answers to the question of the optimal number of hidden neurons/layers as there are many in-house proprietary neural network softwares in the world.

It was found that neural networks with a single hidden layer can do most of the job. It is therefore suggested that you start with a single hidden layer. Neural networks hardly have more than two hidden layers. We hereafter refer to neural networks of only one hidden layer.

There is no theory on the number of hidden neurons. Researchers have thus relied on experimentation and offered a handful of rules of thumb, which can again still be contradicting to one another. Nevertheless, we summarize some of the rules for you to kick off the game. For a neural network with N input neurons and M output neurons, T. Masters suggested a number of $\sqrt{NM}$ hidden neurons. The actual optimal number can still vary between one half and two times the geometrical mean value. D. Baily and D.M. Thompson (J.O. Katz) suggested the number of hidden neurons be 75% (50-300%) of the number of input neurons. C.C. Klimasauskas explicitly linked the number of training data with the number of neurons, suggesting that the number of training facts be at least 5 times that of the weights. The rules can turn out to limit the number of input neurons, which was discussed in step one. We see the interdependence in neural network designs.

The next step concerns output neurons and the error function.

5.6 Error Function

A common error function to be minimized is the squared errors defined in Eq. (5.5). Suppose, for example, the neural network is trained to recommend the user whether to buy or sell yen in the currency exchange market. The output of the neural network can have only one neuron and is therefore a scaler. If the output of the neuron gives a value greater than 0.9, it signals a sell; if it is less than -0.9 then it is a buy. Values between -0.9 and 0.9 render no decision. For the neural network to yield profits, it must be able to predict turning points in the evolving curve of the yen index. Before training, we examine the historical curve of the yen index and assign a 'buy' when the index is at the trough and a 'sell' when it is at the peak. At times between troughs and peaks, the desired (target) neural network outputs are assigned 'no decision'.

During training, weights of the neural network are adjusted in the hope that its outputs (buy, sell, or no decision) over time match the desired ones. The error function is then defined to be the squared differences between the neural network outputs and the desired outputs.

However, we might remove all the 'no decision' target patterns in the error function so that the neural network does not waste weights remembering unimportant fluctuations. All three patterns are however needed in inputs as they make up the continuous history. If the error function is so defined, the neural network is forced to output either buy or sell. The strategy of the user is then changed to buy (sell) yens when the neural network outputs, say, 5 consecutive buys (sells). In this way, the neural network might have detected a falling in the yen price and is predicting a turning over at the fifth buy signal. Patterns less than 5 consecutive buys might simply identify minor troughs which need no attention since they make no profits considering the transaction fees. The actual strategy has to be experimented and depends on the user's portfolio.

The last step in the design is to tune the numerical values of the weights. The error function is a function of the neuron connecting weights whose number is often huge. Furthermore, the function can have lots of local minima in its weight space. A powerful function minimization method capable of finding the global minimum is simulated annealing introduced in Chapter 4 (cf. Section 4.7). Once the neural network is deployed, frequent retraining is beneficial and sometimes mandatory because the important temporal patterns might have changed since the last training.

5.7 Kohonen Self-Organizing Map

Problems are most often solved with greater ease in one way than the other. A properly presented/addressed problem simplifies both the input and output layer of the neural network and hence the architecture. In the rest of the chapter, we introduce and implement a variant of neural networks which is good at clustering multi-dimensional data. Categorized data are often served as inputs to neural networks for pattern recognition.

A Kohonen self-organizing map is a neural network with an input layer and a normally 2-dimensional output layer as shown in Figure 5.4. The unique property of a Kohonen neural network is that it is designed to preserve, during mapping, the topology of the input vectors, which are usually of very high dimensions (namely, have many components), so that, after successful learning, clusters emerge in the (2-dimensional) output layer. Each cluster corresponds to the vectors which are close to one another in the input vector space. We can say that the number of clusters that are formed represents, if the number of neurons in the output layer is large enough, the number of categories there are among the input vectors. We then label the clusters. Later on, a new input vector, when presented to the learned neural network, falls on one of the clus-

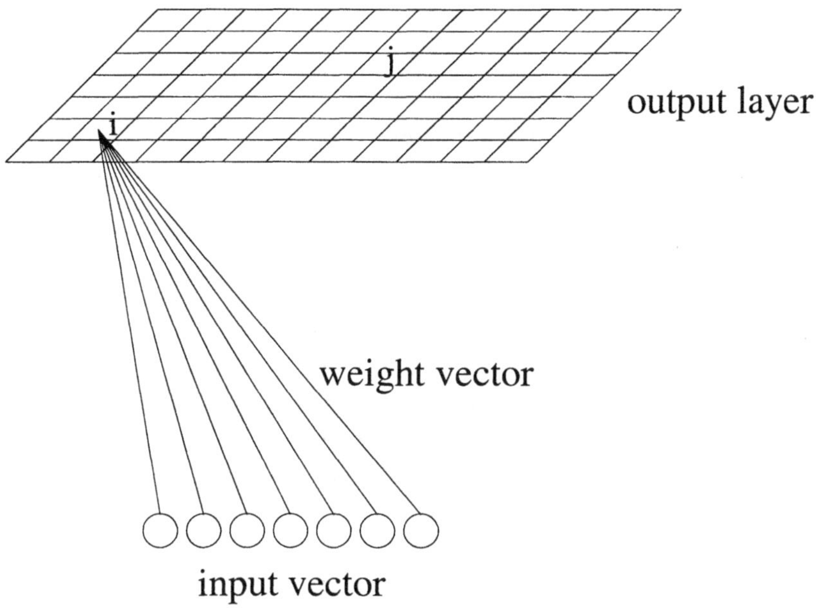

Figure 5.4. Architecture of a Kohonen neural network

ters (labels). A Kohonen neural network therefore works as a classifier. The neural network can also be thought of as performing feature extraction and visualization.

Application of Kohonen's feature map can be found in a broad spectrum of fields. For example, the method was used to classify folk song collections based on the distributions of melodic pitches, intervals, and durations.

In the next section, we describe the learning of the Kohonen neural network which altogether explains the term self-organizing.

5.8 Unsupervised Learning

Learning (or training depending on whose perspective) is an integral part of a neural network, much like a kid learns toward maturing. The learning method described early in this chapter is supervised learning in the sense that desired outputs are given and the neural network's predictions are forced to come close to the desired ones. This goal is accomplished when the values of the weights are so tuned that the error function is minimized.

In contrast, in Kohonen neural networks, given an input vector, the Euclidean distances between the input vector and any of the weight vectors, which has the same dimensions as the input vector, are calculated. The one having the shortest distance is declared the winner and the weight vectors which are in the winner's neighborhood are updated according to the following rule,

$$w_j(k+1) = w_j(k) + \beta(k)C_{ij}(k)(\mathbf{x} - w_j(k)), \tag{5.7}$$

where w_j is the weight vector into neuron j on the output layer, k is the iteration (or time) index. $\beta(k)$ is called learning rate and usually defined as,

$$\beta(k) = \beta_{\text{initial}} \left(\frac{\beta_{\text{final}}}{\beta_{\text{initial}}} \right)^{k/k_{\max}}, \tag{5.8}$$

where $k_{\max}$ is the maximum number of iterations. It is seen that if $\beta_{\text{initial}} = 1.0$ and $\beta_{\text{final}} = 0.05$, the value of β starts from 1.0 and drops as k increases until it becomes 0.05 at $k = k_{\max}$. This is to say the neural network learns less and less harder as time goes on, just like a human does when she grows. $C_{ij}(k)$ defines the neighborhood and in many cases has the following form,

$$C_{ij} = \exp\left[-\frac{1}{2} \left(\frac{|i-j|}{\sigma(k)} \right)^2 \right], \tag{5.9}$$

where $|i - j|$ is the Euclidean distance between neuron i and neuron j in the output layer and $\sigma(k)$ is a linear decreasing function of k. C_{ij} tells the weight vectors closer to the winning weight to move more toward the winner, while those farther away from the winner move less. The effect of the weight updating by Eqs. (5.7), (5.8), and (5.9) is that, after all the input vectors are fed to the neural network, clusters form by themselves in the end of the iteration at $k = k_{\max}$. Kohonen neural networks are thus self-organizing and require no supervision.

5.9 A Clustering Example

Coding of Eqs. (5.7), (5.8), and (5.9) is really simple. To help easily visualize the emergence of clusters on a 2-dimensional plane, we use various instances of the `Color` class in Java programming language to serve as our input data vectors. The `Color` class is based on the 3-component RGB model where `new Color(1.0f,0.0f,0.0f)`, `new Color(0.0f,1.0f,0.0f)`, and `new Color(0.0f,0.0f,1.0f)` generate respectively the 3 primary colors red, green, and blue. Any other colors are obtained by component values between `0.0f` and `1.0f`. For example, `(1.0f,1.0f,0.0f)` gives yellow, `(0.0f,0.0f, 0.0f)` black, `(1.0f,1.0f,1.0f)` white, and so on. Listing 5.1 gives the class responsible for initializing the input data, which are stored in an array of `Neuron` objects defined in Listing 5.2.

```
/*       Sun-Chong Wang
```

```
        TRIUMF
        4004 Wesbrook Mall
        Vancouver, V6T 2A3
        Canada
        e-mail: wangsc@triumf.ca

        DataBase.java initializes the input vectors
        for the Kohonen's self-organizing map */

import java.util.*;

class DataBase extends Observable {
 Neural parent;
 Neuron[] data;  // array of Neuron objects
 final int num_samples = 100;

 public DataBase(Neural parent) {
        super();
        this.parent = parent;
        data = new Neuron[num_samples];
        initializeData();
 }      // end of constructor

 public void DrawMap(Neuron[][] weight, double[][] similarity) {
        parent.plotting.weight = weight;
        parent.plotting.similarity = similarity;
        setChanged();
        notifyObservers(parent.plotting);
 }      // end of method

 private void initializeData() {
        Random rand = new Random();
        rand.setSeed(1L);

        // 100 different colors
        for (int i=0; i<num_samples; i++) {
            data[i] = new Neuron();
            data[i].x = rand.nextDouble();
            data[i].y = rand.nextDouble();
            data[i].z = rand.nextDouble();
        }
        // six (or eight) different colors
        for (int i=0; i<num_samples/2; i++) {
            data[i] = new Neuron();
            data[i].x = 1.0;           // red
            data[i].y = 0.0;
            data[i].z = 0.0;
        }
        for (int i=50; i<60; i++) {
            data[i] = new Neuron();
            data[i].x = 0.0;           // green
            data[i].y = 1.0;
            data[i].z = 0.0;
        }
        for (int i=60; i<70; i++) {
            data[i] = new Neuron();
            data[i].x = 0.0;           // blue
            data[i].y = 0.0;
            data[i].z = 1.0;
        }
        for (int i=70; i<80; i++) {
            data[i] = new Neuron();
            data[i].x = 1.0;           // yellow
            data[i].y = 1.0;
            data[i].z = 0.0;
        }
        for (int i=80; i<85; i++) {
```

```
            data[i] = new Neuron();
            data[i].x = 0.0;            // cyan
            data[i].y = 1.0;
            data[i].z = 1.0;
        }
        for (int i=85; i<90; i++) {
            data[i] = new Neuron();
            data[i].x = 1.0;            // magenta
            data[i].y = 0.0;
            data[i].z = 1.0;
        }
        for (int i=90; i<95; i++) {
            data[i] = new Neuron();
            data[i].x = 0.0;            // black
            data[i].y = 0.0;
            data[i].z = 0.0;
        }
        for (int i=95; i<100; i++) {
            data[i] = new Neuron();
            data[i].x = 1.0;            // white
            data[i].y = 1.0;
            data[i].z = 1.0;
        }
 }      // end of initializeData
}       // end of class DataBase
```

Listing 5.1 DataBase.java

```
/*      Sun-Chong Wang
        TRIUMF
        4004 Wesbrook Mall
        Vancouver, V6T 2A3
        Canada
        e-mail: wangsc@triumf.ca

        Neuron.java defines the class to hold the input
        vector as well as the weights */

public class Neuron {
 public double x,y,z;

 public Neuron() {
        x = 0.0;
        y = 0.0;
        z = 0.0;
 }      // end of constructor
}       // end of class Neuron
```

Listing 5.2 Neuron.java

In the first illustration of clustering, we prepare for six input colors. The weights are assigned random colors. In the beginning of the learning, you thus see colorful dots randomly distributed across the canvas. As time goes on, similar colors aggregate and finally in the end (after 10,000 iterations) six blocks of distinct colors are really formed as shown in Figure 5.5. The network is performing clustering !

Note that when you run the learning again (with different initial random weights by different seeds to the random number generator), six clusters still form but their locations on the map may change. This is because of the random initial weights. The implication is that the similarity between adjacent clusters is not necessarily higher than that of disjoint clusters. Segregation patterns depend on initialized values of the weights. We therefore write, in `SOM.java`, the

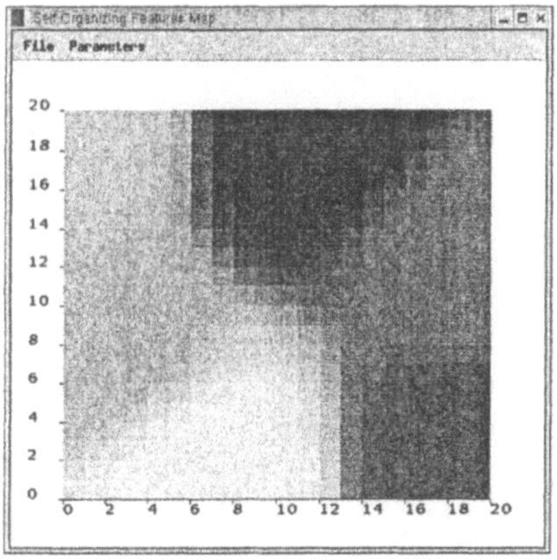

Figure 5.5. Six clusters resulting from six input colors

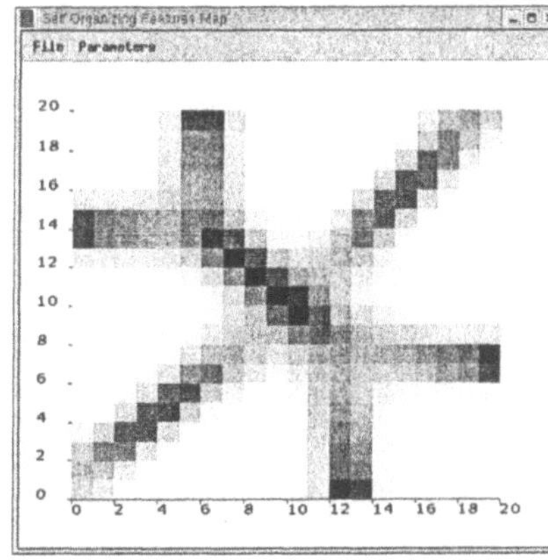

Figure 5.6. Similarity plot of the clusters in Figure 5.5

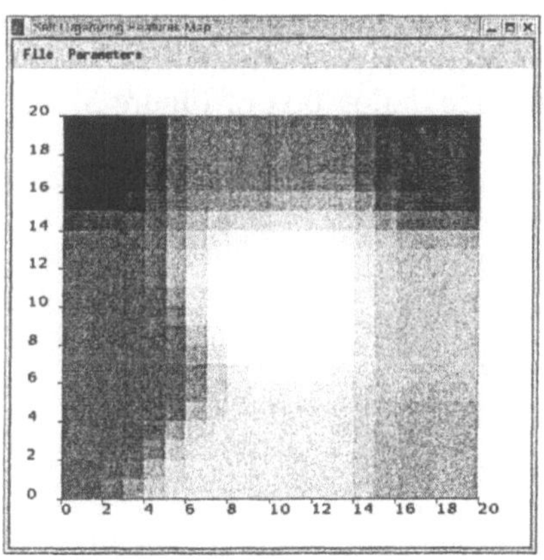

Figure 5.7. Clustering with eight input colors

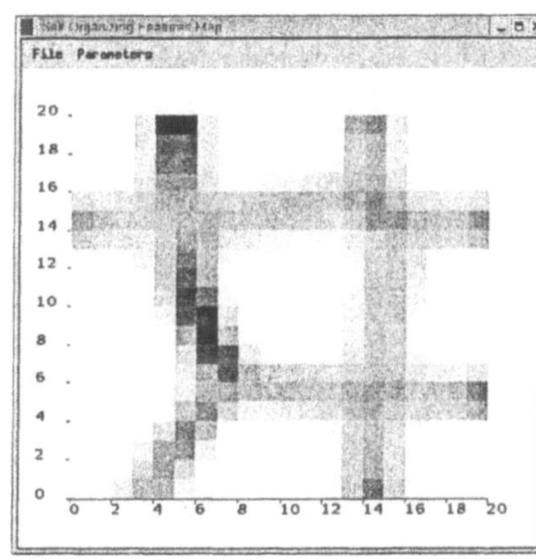

Figure 5.8. Similarity plot of Figure 5.7

`Similarity()` method which calculates the average distance between neighboring weights. Darker colors are assigned to larger average distances in the plotting class `Plotter.java`. The similarity plot associated with Figure 5.5 is shown in Figure 5.6. It is seen there that clusters are isolated by dark ridges.

In the second illustration, we input eight colors to the same program. Results of the feature map and similarity map are shown in Figures 5.7 and 5.8.

The source code for the learning is given in Listing 5.3. Now let's increase the number of different input colors to 100. The resulting maps are shown in

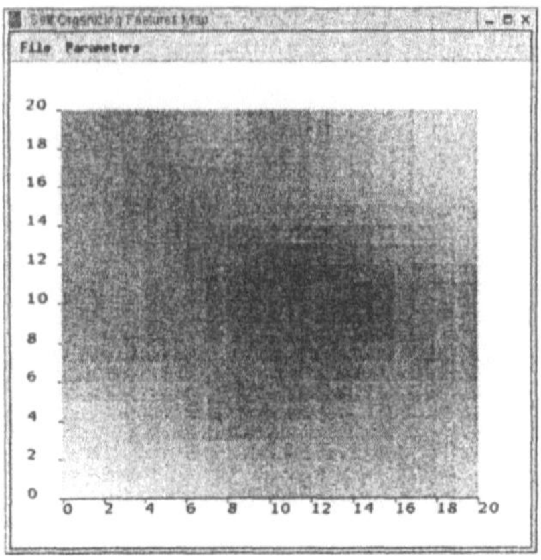

Figure 5.9. Clustering 100 different input colors

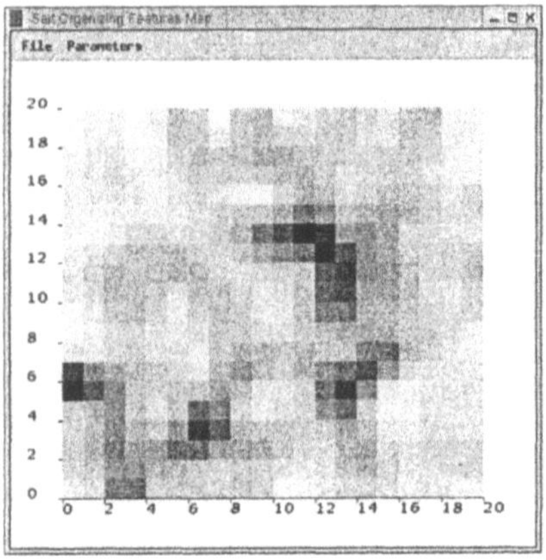

Figure 5.10. Similarity plot of Figure 5.9

Figures 5.9 and 5.10. It is noticed that a grid of 20 by 20 output neurons might be insufficient as evidenced by the blurring boundaries in the similarity map of Figure 5.10. We then increase the number of output layer neurons to 40 by 40 (Figures 5.11 and 5.12) through the user interface dialog box of Figure 5.13.

```
/*      Sun-Chong Wang
        TRIUMF
        4004 Wesbrook Mall
        Vancouver, V6T 2A3
        Canada
        e-mail: wangsc@triumf.ca

        SOM.java codes the unsupervised learning algorithm of
        the self-organizing map: Eqs. (5.7), (5.8), (5.9) */

import java.lang.*;
import java.util.*;
import java.util.Random;

public class SOM implements Runnable { // a thread
 Random rand;

 int XSize, YSize;
 int iTime, ifreq;
 int num_samples;
 // initial/final beta/sigma
 double beta_i, beta_f, sigma_i, sigma_f;
 double[][] distance,similarity;
 Neuron[][] weight;
 Neuron[] data;

 SOMDialog parent;

 public SOM(SOMDialog parent) {
        this.parent = parent;

        rand     = new Random();
        System.out.println("random number = " + rand.nextFloat());

        XSize = 0;
```

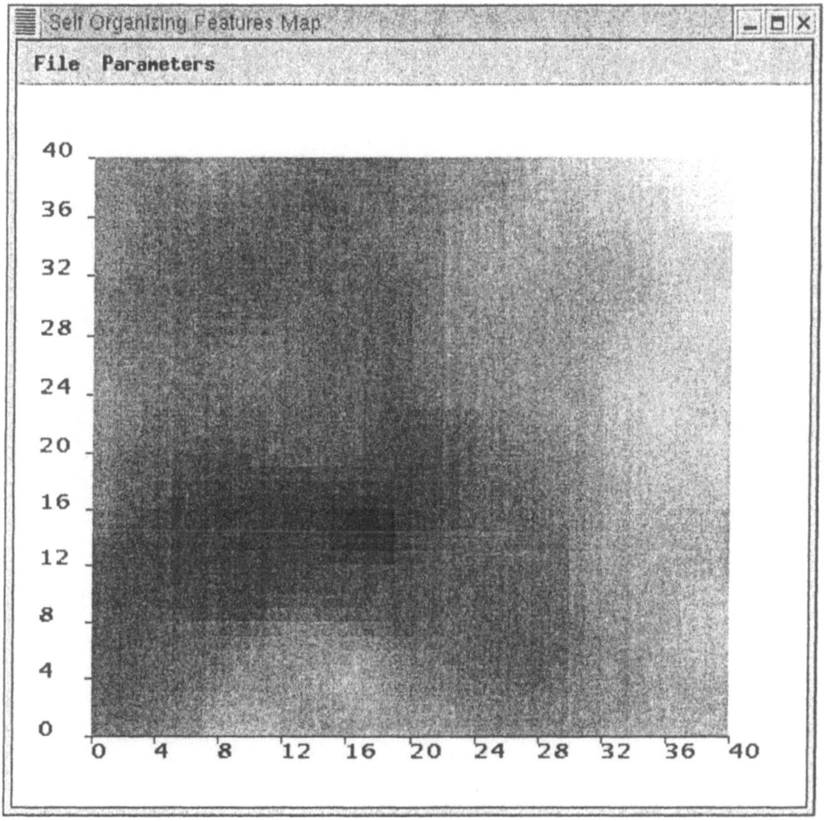

Figure 5.11. 100 input colors and 4 times of output neurons

```
        YSize = 0;
        iTime = 10000;
        ifreq = 100;
        beta_i = 1.0;
        beta_f = 0.01;
        sigma_i = (XSize+YSize)/2.0/2.0;
        sigma_f = 1.0;
}       // end of SOM class constructor

public void run() { // executed by a thread
        for (int i=0; i<iTime; i++) {
            Learning(i);
            if (i%ifreq == 0) {
               Similarity();
               parent.parent.db.DrawMap(weight,similarity);
            }  // db is an instance of the DataBase class
            parent.jbar.setValue(Math.round(((float)i)/
                                  ((float)iTime)*100.0f));
        }
}

public void Setup(int XSize, int YSize) {
```

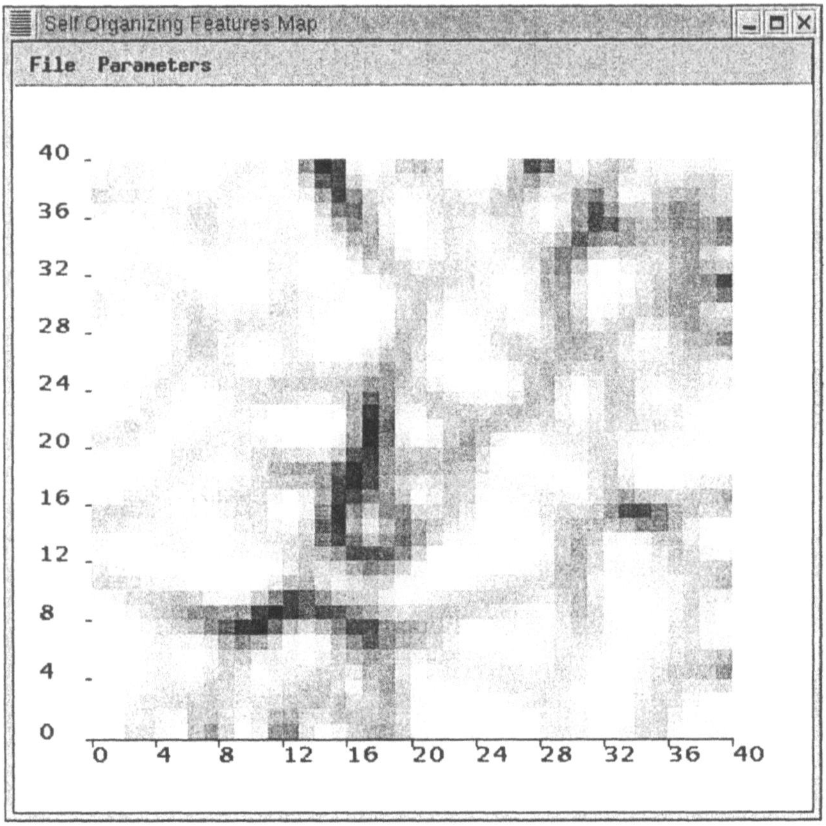

Figure 5.12. Similarity plot of Figure 5.11

```
if (XSize != this.XSize || YSize != this.YSize) {
   this.XSize = XSize;
   this.YSize = YSize;
   distance = new double[XSize][YSize];
   weight = new Neuron[XSize][YSize];  // array of objects
   for (int i=0; i<XSize; i++) {
       for (int j=0; j<YSize; j++) {
           weight[i][j] = new Neuron();// call the constructor
       }
   }
   similarity = new double[XSize][YSize];
   // average radius
   sigma_i = (XSize+YSize)/2.0/2.0;
} // end if

rand.setSeed(2L);
for (int i=0; i<XSize; i++) {
    for (int j=0; j<YSize; j++) {
        weight[i][j].x = rand.nextDouble();
        weight[i][j].y = rand.nextDouble();
        weight[i][j].z = rand.nextDouble();
```

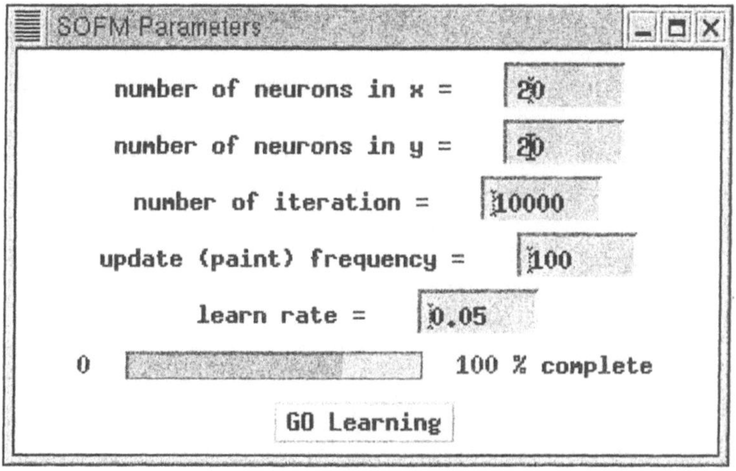

Figure 5.13. User interface for the SOM application

```
            }
        }
}       // end of Setup

private void Learning(int itime) {
        int i_win=0,j_win=0;
        double exponent,beta,sigma,adjust,distance_ij,dmin;

        // randomly picks up an input vector
        int k = rand.nextInt(num_samples);

        dmin = Double.MAX_VALUE;
        // calculates the distance between the input vector and
        // the weight vector.  i,j are the coordinates of the weight
        // on the 2-dimensional output layer
        for (int i=0; i<XSize; i++) {
            for (int j=0; j<YSize; j++) {
                distance[i][j] = (data[k].x - weight[i][j].x)*
                                 (data[k].x - weight[i][j].x)+
                                 (data[k].y - weight[i][j].y)*
                                 (data[k].y - weight[i][j].y)+
                                 (data[k].z - weight[i][j].z)*
                                 (data[k].z - weight[i][j].z);
                if (distance[i][j] < dmin) {
                   dmin = distance[i][j];
                   // keeps the coordinates of the winner
                   i_win = i;
                   j_win = j;
                }
            }
        }

        exponent = itime/((double)(iTime-1.0));
        beta = beta_i*Math.pow(beta_f/beta_i,exponent); // Eq. (5.8)
        sigma = sigma_i*Math.pow(sigma_f/sigma_i,exponent);

        for (int i=0; i<XSize; i++) {
            for (int j=0; j<YSize; j++) {
```

```
                // distance to the winning weight
                distance_ij = (i-i_win)*(i-i_win)+(j-j_win)*(j-j_win);
                // Eq. (5.9)
                adjust = beta*Math.exp(-distance_ij/sigma/sigma/2.0);
                // the update rule: Eq. (5.7)
                weight[i][j].x += adjust*(data[k].x-weight[i][j].x);
                weight[i][j].y += adjust*(data[k].y-weight[i][j].y);
                weight[i][j].z += adjust*(data[k].z-weight[i][j].z);
            }
        }
}       // end of Learning

// average distance between the weight vectors
// to serve as a measure of similarity
private void Similarity() {
        double max;

        for (int i=1; i<XSize-1; i++) {
            for (int j=1; j<YSize-1; j++) {
                similarity[i][j] = (w_distance(i,j,i-1,j)+
                                    w_distance(i,j,i+1,j)+
                                    w_distance(i,j,i,j-1)+
                                    w_distance(i,j,i,j+1))/4.0;
            }
        }

        // boundary cells
        for (int j=1; j<YSize-1; j++) {
            similarity[0][j] = (w_distance(0,j,0,j+1)+
                                w_distance(0,j,0,j-1)+
                                w_distance(0,j,1,j))/3.0;
            similarity[XSize-1][j] =
                      (w_distance(XSize-1,j,XSize-1,j+1)+
                      w_distance(XSize-1,j,XSize-1,j-1)+
                      w_distance(XSize-1,j,XSize-2,j))/3.0;
        }
        for (int i=1; i<XSize-1; i++) {
            similarity[i][0] = (w_distance(i,0,i+1,0)+
                                w_distance(i,0,i-1,0)+
                                w_distance(i,0,i,1))/3.0;
            similarity[i][YSize-1] =
                      (w_distance(i,YSize-1,i+1,YSize-1)+
                      w_distance(i,YSize-1,i-1,YSize-1)+
                      w_distance(i,YSize-1,i,YSize-2))/3.0;
        }

        // corner cells
        similarity[0][0] =
           (w_distance(0,0,0,1)+w_distance(0,0,1,0))/2.0;
        similarity[0][YSize-1] =(w_distance(0,YSize-1,0,YSize-2)+
                                 w_distance(0,YSize-1,1,YSize-1))/2.0;
        similarity[XSize-1][YSize-1] = (w_distance(XSize-1,YSize-1,
                                        XSize-2,YSize-1)+
                                        w_distance(XSize-1,YSize-1,
                                        XSize-1,YSize-2))/2.0;
        similarity[XSize-1][0] = (w_distance(XSize-1,0,XSize-2,0)+
                                 w_distance(XSize-1,0,XSize-1,1))/2.0;

        max = 0.0;
        for (int i=0; i<XSize; i++) {
            for (int j=0; j<YSize; j++) {
                if (similarity[i][j] > max) max = similarity[i][j];
            }
        }
```

```
        // normalizing (for black and white plotting)
        for (int i=0; i<XSize; i++) {
            for (int j=0; j<YSize; j++) {
                similarity[i][j] /= max;
            }
        }
}       // end of Similarity

private double w_distance(int i, int j, int m, int n) {
        double tmp;
        tmp = (weight[i][j].x-weight[m][n].x)*
              (weight[i][j].x-weight[m][n].x)+
              (weight[i][j].y-weight[m][n].y)*
              (weight[i][j].y-weight[m][n].y)+
              (weight[i][j].z-weight[m][n].z)*
              (weight[i][j].z-weight[m][n].z);
        return tmp;
}       // end of w_distance
}       // end of SOM class
```

Listing 5.3 SOM.java

5.10 Summary

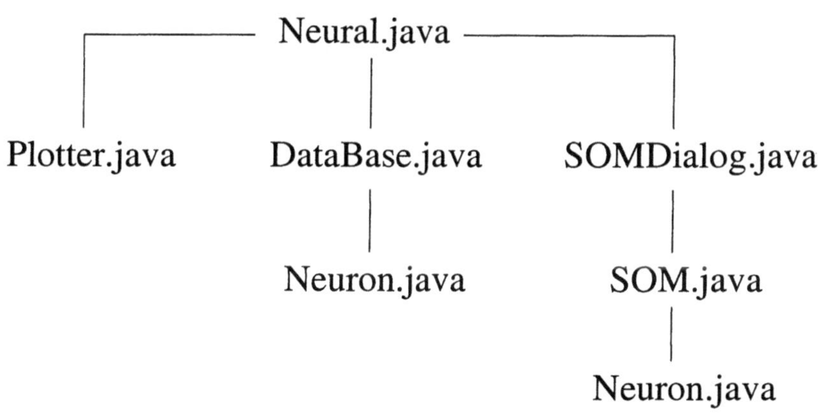

Figure 5.14. Source files for the Kohonen self-organizing map in this chapter

`Neural.java`, containing the `main()` method, is easily written using `Spin.java` of Chapter 4 as a template. `Plotter.java`, extending `Canvas` and implementing `Observer`, is similar to the one in the appendix. `SOMDialog.java`, a dialog box for user interaction, can be obtained by modifying the dialog box class, `SADialog.java`, in Chapter 4.

We implemented a Kohonen self-organizing map. Vectors of colors were input to the Kohonen neural network, and grouped into clusters on its own on the 2-dimensional output layer in the end of the learning.

We laid out steps when designing a neural network. A recurrent neural network architecture which has 'memory' neurons was introduced. Economic time series prediction was used as an example throughout the designing steps.

A neural network, after training, is able of generalizing the patterns embedded in the training data set. However, it is not expected to detect patterns that do not exist in the training data. A neural network can become more powerful when its predicting/recognizing capability is combined with adaptivity, which is the subject of the next chapter.

5.11 References and Further Reading

An introduction to self-organizing map is, T. Kohonen, "Self-organizing Maps", 2nd ed. Springer-Verlag, Berlin (1997)

Two textbooks on neural networks are, C.M. Bishop, "Neural Networks for Pattern Recognition", Oxford University Press, Oxford (1995), and B.D.Ripley, "Pattern Recognition and Neural Networks", Cambridge University Press, Cambridge (1996)

Numerous resources on neural networks can be found in the on-line FAQ located at, `ftp://ftp.sas.com/pub/neural/FAQ.html`

Chapter 6

GENETIC ALGORITHM

Organisms are one of the most wonderful systems in the world. Like the method of the last chapter, we introduce here another powerful problem solving technique inspired from biology. Genetic algorithm, just like simulated annealing, is suitable to both combinatorial and numerical optimizations. They find wide applications in different research fields, such as management, engineering, industrial design, and so forth.

6.1 Evolution

Living systems evolve and adapt in harsh environments in order to survive according to the theory of natural selection. In this evolutionary view, individuals with some traits which have proven fitter to the environment fare better and thus have higher chances of propagating, via reproduction, the fitter characteristics to the next generation. On the other hand, those with disadvantageous or harmful traits are disfavored and would eventually cease to exist in the population.

It is known in molecular biology that hereditary information is encoded and stored in double-stranded helices called DNA (Deoxyribonucleic Acid). DNA molecules are organized as chromosomes. Elemental information carriers in a DNA molecule are genes and chromosomal DNA forms the genome of the organism. For sexual reproduction, threads of replicated chromosomes from one parent pair with corresponding ones from the other parent in forming a germ cell. Traits of parents are inherited.

In the genetic algorithm, we compare the maximum (or minimum) of an objective function to the fittest to an environment. Parameters of the objective function are stored in a 1-dimensional array which is likened to a chromosome. We prepare an initial population of parents (i.e., a pool of parameter arrays) and then let them evolve under the process of natural selection.

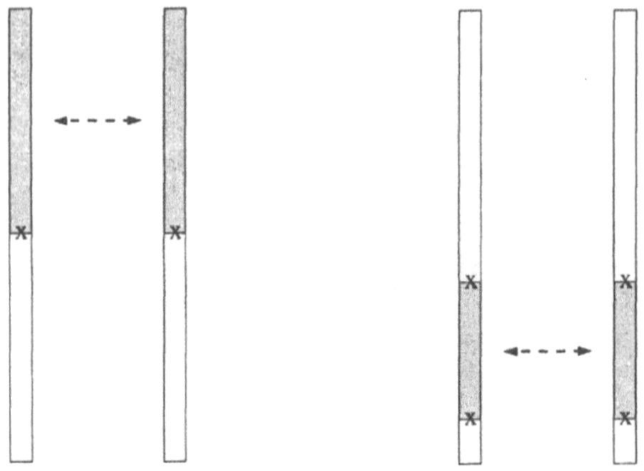

(a) (b)

Figure 6.1. Crossover operation

6.2 Crossover

During evolution, traits of the parents are mixed in the hope that good traits are preserved and passed to filial generations. This mechanism bolsters long lasting prosperity of the lineage. In function optimization, mixing can be achieved by *crossover* operation (meiosis in biology). In the operation, the array storing the parameters is cut into halves. The first half of the array from one parent is then recombined to the second half of the array from the other parent. More general crossover operations are shown in Figure 6.1. In Figure 6.1 (a), a breakup point 'x' of the array is randomly selected. Corresponding pieces of the cut arrays are then exchanged. For the purpose of illustration, we take the famous traveling salesman problem as an example. Suppose the salesman is going to travel 12 cities cyclically. He likes to plan the order of traveling so that the total distance traveled is the shortest, and therefore the most economical. To solve the problem by the method of genetic algorithm, we prepare an initial population of, say, 100 trial traveling orders stored in 100 arrays. The elements of each array are integers between 0 and 11 (inclusive), representing the 12 cities. Note that, since every city has to be visited, integers do not repeat in the array.

Various ways of crossing over can be experimented. For example, an array can be cut at two breakup points, with the middle piece exchanged between the parents, as in Figure 6.1 (b). The breakup points are randomly selected, so even same pairs of parents will produce different children. The action of crossover and recombination therefore ushers genetic diversity into the population. We implement this type of crossover operation in the code of this chapter.

Crossover operations can be thought of as probing global features of the landscape in the quest for the global minimum. For example, if 3-8-2 and 9-5-1-6 are already two good sequences, it is hoped that after crossover between the parents they are combined to yield a better solution to the problem.

In real implementation, after parents are selected from the population, not every pair of them is crossed over. A probability, P_c, called crossover rate determines the frequency. Best values of P_c vary from case to case. We can guess a value of 0.7 to start.

6.3 Mutation

Mutant genes in a living system can sometimes lead to malfunction of the cell, causing trouble to the system. In other cases, however, they can turn out to endow the system with extra-ordinary capabilities, giving birth to a very competitive entity in the population.

Mutation occurs on DNA itself. In our traveling salesman example, it can be realized by swapping two cities randomly selected from the array, resulting in a new traveling sequence. In applications such as the Ising model in Chapter 4, mutation can be performed by randomly choosing a spin from the lattice sites and then reversing its sign. In late stages of evolution, mutation can be thought of as a way of preventing loss of diversity in the population. It therefore ensures that evolution is not trapped in local minima. Similar to the crossover rate, we can define a mutation rate, P_m, which determines the proportion of the population undergoing mutation. A starting value of $P_m = 0.1$ can be tried.

Several different forms of mutation can be defined and operated at the same time to accelerate exploration of the configuration space. For example, it could be found that, in the traveling salesman problem, the genetic algorithm often returns solution arrays of $(\cdots, N_{i-1}, N_{i+1}, N_i, \cdots)$ whereas we know somehow that the true solution should be $(\cdots, N_{i-1}, N_i, N_{i+1}, \cdots)$. The program is trapped in local minima ! Based on the observation, we can then devise a mutation operation which randomly picks up two consecutive cities, and then reverses their order. While a drug cures an ailment, two different drugs applied at the same time can become a poison. A caveat is therefore that we may want to limit the dosage of joint mutations.

Relative merits of crossover and mutation are case dependent. When developing the methodology, We need to adjust the weights, P_c, P_{m_1}, P_{m_2}, $P_{m_3}, \cdots$, for rapid convergence to the minimum. The Java application of this chapter is equipped with a pop-up dialog box holding parameters such as P_c and P_m to facilitate experimenting.

6.4 Selection

We have discussed the genetic operators on the arrays (parents) but left out an important premise: how do we select parents? There are as many different implementations of selection as many operations of crossover and mutation. Nevertheless, the goal is common; it is for fitter individuals to survive and gradually dominate the population. To achieve this, we can, for example, assign to the individual a score which is related to its performance on the objective function. The score then serves as a measure of rate at which the individual is selected as parent in repeated trials. This can best be understood by an analogy to a roulette wheel in a casino. If the edge of the wheel is unevenly partitioned, the wider the partition, the more likely the partition is visited as the revolving wheel comes to a stop.

The selection problem now becomes a scoring one. There can be many ways of transforming objective function values to scores. Most often, some controlling parameters are introduced into the transformation in such a way that seemingly fitter individuals do not prematurely predominate the population. On the other hand, favorable individuals should be properly promoted. For example, one can map the objective function values into probabilities by the use of the Boltzmann weight introduced in Chapter 4. In the weighting, a controlling parameter called 'temperature' delineates the relative abundance of the state of the system in the ensemble. How to balance the two contradicting factors (prematurity aversion versus favorable promotion) is in most cases as much an art as a science. As a result, a heuristic transformation is key to rapid and robust convergence.

In the following, we introduce a simple transformation, namely, selection by *tournament*. In this scheme, firstly, two contenders are randomly drawn from the population. The one who outperforms the other (returns a shorter distance in the traveling salesman problem) wins the tournament and is chosen as one of the parents. Next, a second pair of players are randomly selected from the same population pool excluding the previous winning contender. Then the winner is chosen as the second parent. Next, the two chosen parents mate to produce two children by crossover followed by mutation. The tournament method makes sure that fitter individuals get higher chances of being selected, promoting their advantageous ingredients to the next generation. Weak individuals still have chances of reproducing if two weak players are drawn to match. They keep the diversity of the population pool. However, in some particular cases where diversity is an issue of less importance, we can device the tournament rule so that a contender who wins the first match is only qualified. A final round of match is then held between two qualified players. The winner of the final game is selected as one of the parents. The rule of 'win 2 games in a row' enforces that really tough individuals are selected. There might be as many different

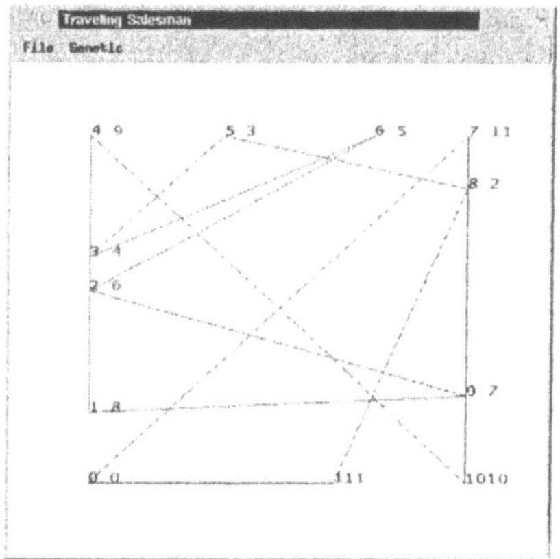

Figure 6.2. City (blue integer) coordinates and an initial traveling sequence (red integers)

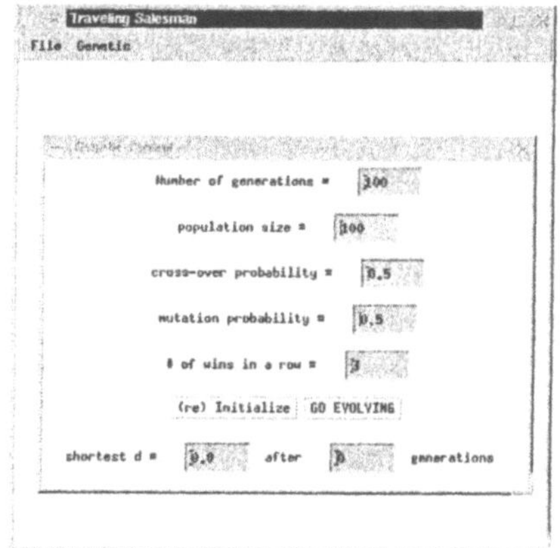

Figure 6.3. The dialog box holding the program parameters and outputting the shortest distance so far

rules as the programmer can imagine. Different rules in the tournament method play the same role as the different controlling parameters in the aforementioned transformation method. A feature of the tournament mechanism, besides its simplicity, is that it is amenable to parallel computing.

6.5 Traveling Salesman Problem

We demonstrate traveling salesman problem using genetic algorithm. On grounds of pedagogy, the 12 cities are located along the four edges of a regular rectangle, as shown by the blue numbers in Figure 6.2. In this case, the solutions which minimize the total traveling distance are known to be either (0, 1, 2, $\cdots$, 11, 0) or (0, 11, 10, $\cdots$, 1, 0) and the shortest distance is 4 times the length of the edge, which is 80 in this example.

In the beginning, we prepare a pool of arrays, each of which is formed by randomly assigning one of the 12 integers to the array element. Note that an integer appears exactly once in the array since a city is visited only once. The size of the pool is a parameter that a user can change through the dialog box. The journey starts from city 0. An initial traveling path is shown by the sequence of red integers in Figure 6.2. Summing up the length of each red line gives us the total traveling distance.

The class `Genetic` (Listing 6.1) then proceeds with selecting parents using the tournament method. Our implementation of the method accepts an integer argument prescribing how many wins in a row in order for the con-

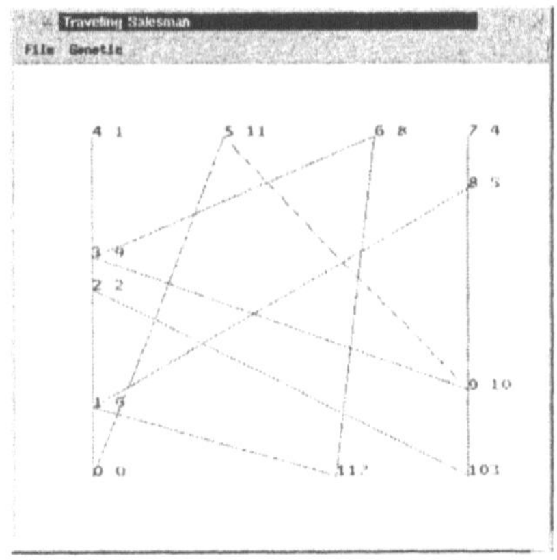

Figure 6.4. The best solution at generation 3

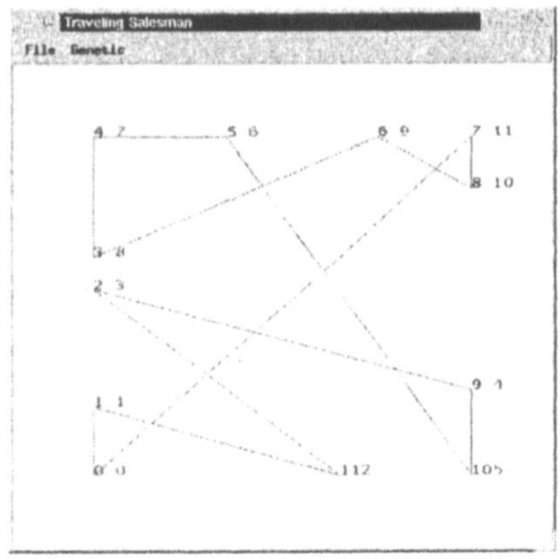

Figure 6.5. The best solution at generation 6

tending array to be selected as parent. Note the recursive call to itself of the `tournament()` method.[1] Two successful parents are then crossed over and mutated to form two children. The children's traveling distances are calculated. After an enough number of children (equal to the size of the population) are generated, the program finds and plots the best traveling sequence on the canvas window and also writes out its distance to the dialog box. The program then continues to the next generation. The user can witness how the algorithm converges to the best solution along the course of the evolution. When the pre-set number of generations is reached, the program quits.

```
/*      Sun-Chong Wang
        TRIUMF
        4004 Wesbrook Mall
        Vancouver, V6T 2A3
        Canada
        e-mail: wangsc@triumf.ca

        Genetic.java encapsulates the genetic operations
        for the genetic algorithm */

import java.lang.*;
import java.util.*;
import java.util.Random;

public class Genetic {
 int ngens;               // number of generations
 int popsize;             // number of individuals in population
 double p_cross;          // probability of crossover
 double p_mutate;         // probability of mutation
```

[1]Recursion can save lots of lines in the code. However, since each call creates its own stack for variables/objects, if improperly designed, the method can drain the system of memory quickly. Care has to be taken when programming recursion in Java.

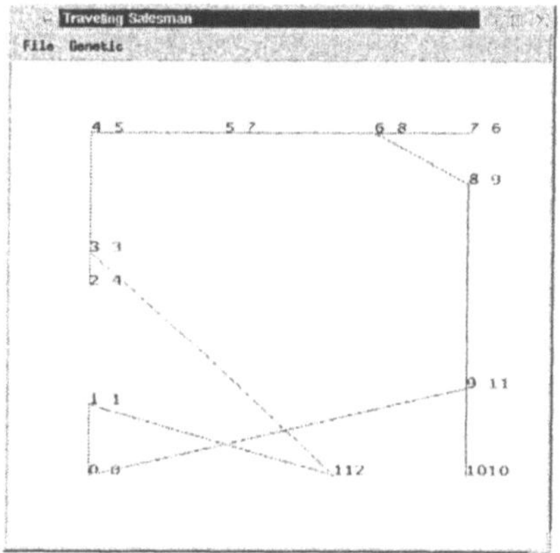

Figure 6.6. The best solution at generation 9

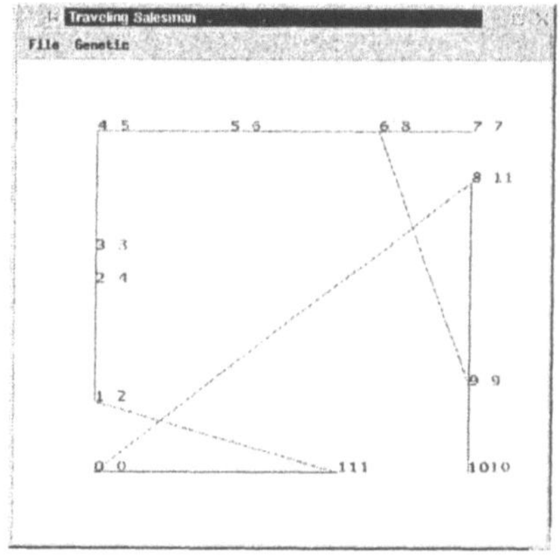

Figure 6.7. The best solution at generation 12

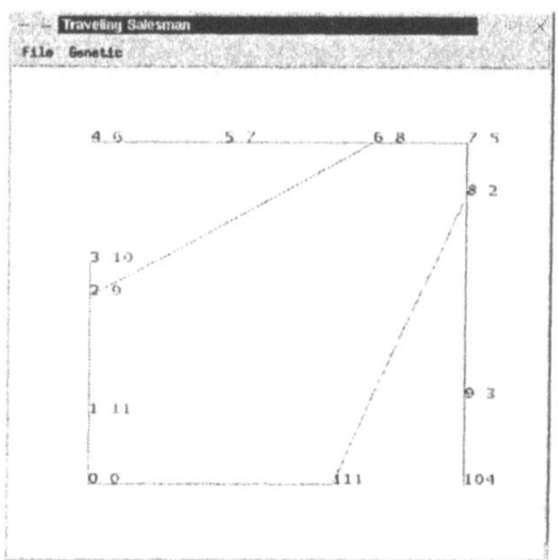

Figure 6.8. The best solution at generation 15

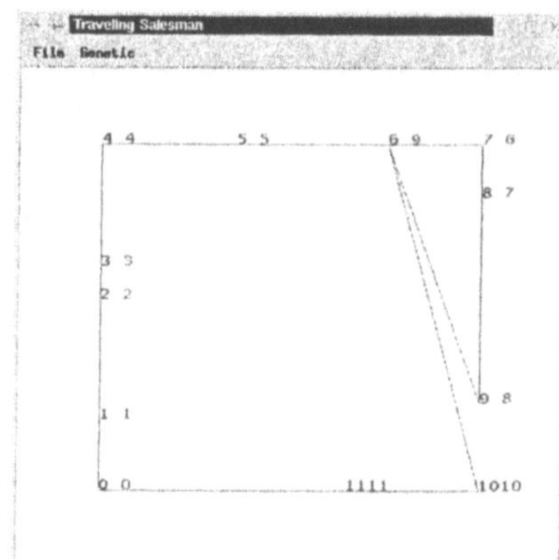

Figure 6.9. The best solution at generation 18

```
double fquit;           // quit if function reduced to this amount
final int nvars = 12;   // number of variables (cities)
double[] schedule;      // the array for the traveling sequence
double[] best;          // best soultion
int[] choices;          // parent index
double[] fvals;         // function values
double[] newfvals;      // children's function values
double[][] pool1;       // population
double[][] pool2;       // population
int nchoices;           // number of choices
int parent1, parent2;   // indices of the selected parents
int crosspt1,crosspt2;// crossover point of the array
int in_a_row;           // number of wins in a row
double[][] city;        // vectors to store x, y coordinates of cities
```

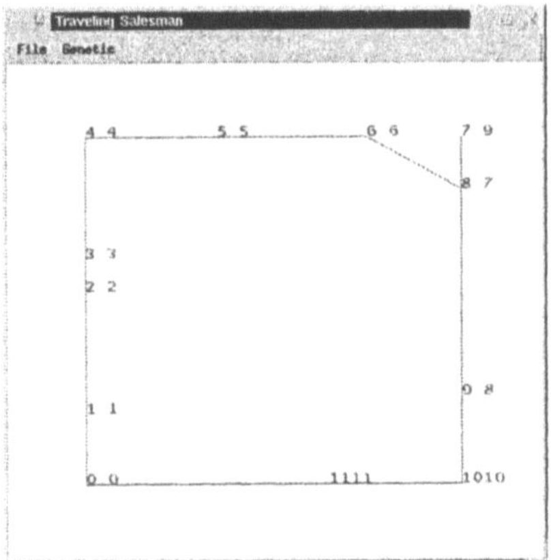

Figure 6.10. The best solution at generation 21

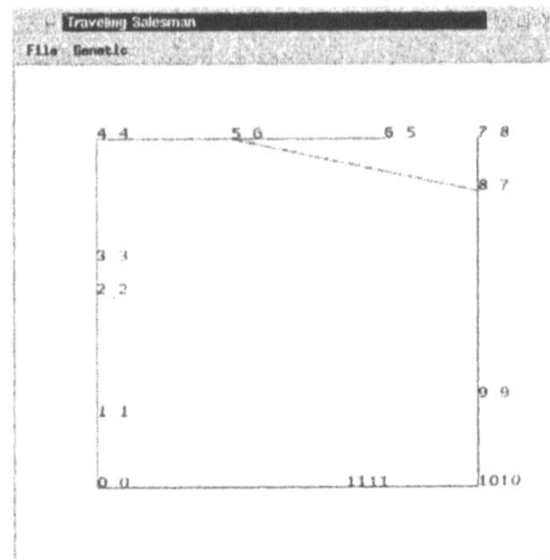

Figure 6.11. The best solution at generation 24

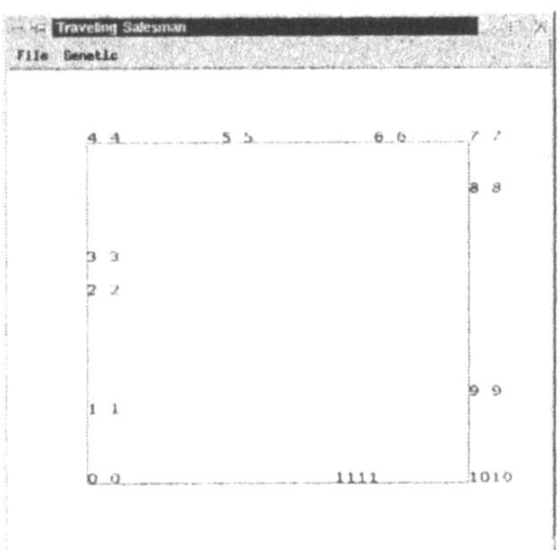

Figure 6.12. The best solution at generation 27

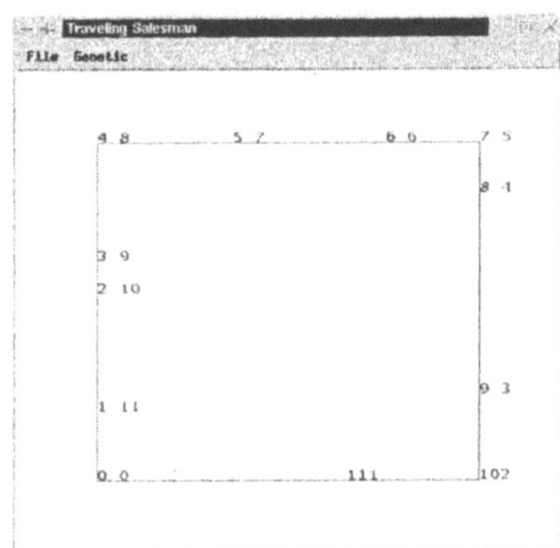

Figure 6.13. The other possible global minimum solution

```
double bestfval;        // shortest distance
int best_i;             // index
double[][] oldpop, newpop;

Random rand;
GADialog parent;

public Genetic(GADialog parent) {
        this.parent    = parent;
        rand           = new Random();
        System.out.println("random number = " + rand.nextFloat());
        ngens = 100;
        popsize = 0;      // between 120 and 600
        p_cross = 0.5;
        p_mutate = 0.5;
```

```
        in_a_row = 3;
        fquit = 80.0;      // the lowest value is 80.0

        city = new double[nvars][2];
        schedule = new double[nvars];
        best = new double[nvars];
        // initialize the sequence, will be randomized later
        for (int i=0; i<nvars; i++) schedule[i] = (double) i;

        // city coordinates
        city[0][0] = 0.0;              city[1][0] = 0.0;
        city[0][1] = 0.0;              city[1][1] = 4.0;

        city[2][0] = 0.0;              city[3][0] = 0.0;
        city[2][1] = 11.0;             city[3][1] = 13.0;

        city[4][0] = 0.0;              city[5][0] = 7.0;
        city[4][1] = 20.0;             city[5][1] = 20.0;

        city[6][0] = 15.0;             city[7][0] = 20.0;
        city[6][1] = 20.0;             city[7][1] = 20.0;

        city[8][0] = 20.0;             city[9][0] = 20.0;
        city[8][1] = 17.0;             city[9][1] = 5.0;

        city[10][0] = 20.0;            city[11][0] = 13.0;
        city[10][1] = 0.0;             city[11][1] = 0.0;
}       // end of Genetic class constructor

public void initialize(int size) {
        if (this.popsize != size) {
           this.popsize    = size;
           choices         = new int[size];
           fvals           = new double[size];
           newfvals        = new double[size];
           pool1           = new double[size][nvars];
           pool2           = new double[size][nvars];
        }

        bestfval = Double.MAX_VALUE;
        best_i = 0;         // Safety only
        // randomize to prepare for the first generation
        for (int i=0; i<popsize; i++) {
            shake(schedule, pool1[i]);
            fvals[i] = Distance(pool1[i]);
        }
        // need this for the first plot
        for (int j=0; j<nvars; j++) best[j] = pool1[0][j];
        bestfval = fvals[0];

        oldpop = pool1;
        newpop = pool2;
}       // end of method initialize

public double[] getSchedule() {
        return best;
}

// calculate the distance
private double Distance(double[] schedule) {
        int i,j;
        double Ax,Ay,Bx=0.0,By=0.0,D;

        D = 0.0;

        Ax = city[0][0];
```

```
        Ay = city[0][1];

        for (j=1; j<schedule.length; j++) {
            for (i=1; i<schedule.length; i++) {
                if ((int) schedule[i] == j) {
                   Bx = city[i][0];
                   By = city[i][1];
                   D += Math.sqrt((Bx-Ax)*(Bx-Ax)+(By-Ay)*(By-Ay));
                   Ax = Bx;
                   Ay = By;
                }
            }
        }
        D += Math.sqrt((Bx-city[0][0])*(Bx-city[0][0])+
                       (By-city[0][1])*(By-city[0][1]));
        return D;
}       // end of Distance

public double[] Go() {
        return genetic();
}

private double[] genetic() {  // the genetic algorithm
        int i, j, k, n_cross;  // for each generation
        boolean first_child, improved;
        double[][] temppop;

        for (i=0; i<popsize; i++) choices[i] = i;
        nchoices = popsize;
        n_cross = (int) (p_cross * popsize);
        first_child = true;
        improved = false;

        for (i=0; i<popsize; i++) {

            if (first_child) {
               parent1 = pick_parent(in_a_row, nchoices);
               parent2 = pick_parent(in_a_row, nchoices);
            } // parents are selected by the tournament method
               // as described in the text

            // crossover
            if (n_cross-- > 0)
               reproduce(first_child, newpop[i], oldpop);
            else if (first_child)
               for (k=0;k<nvars;k++) newpop[i][k]=oldpop[parent1][k];
            else
               for (k=0;k<nvars;k++) newpop[i][k]=oldpop[parent2][k];

            // mutation
            if (p_mutate > 0.0) mutate(newpop[i]);

            newfvals[i] = Distance(newpop[i]);

            if (newfvals[i] < bestfval) {
               bestfval = newfvals[i];
               best_i = i;
               improved = true;
            }

            if (newfvals[i] <= fquit) break;

            first_child = !first_child;
        }  // for i loop

        if (improved) for (k=0;k<nvars;k++) best[k]=newpop[best_i][k];
        for (k=0; k<popsize; k++) fvals[k] = newfvals[k];
```

```
        temppop = oldpop;
        oldpop = newpop;
        newpop = temppop;

        return best;
}       // end of method genetic

private int pick_parent(int in_a_row, int nchoices) {
        int ipick = 0;
        double min = Double.MAX_VALUE;
        int[] itmp = new int[in_a_row];

        tournament(in_a_row, nchoices, itmp);

        for (int i=0; i<itmp.length; i++) {
            if (fvals[itmp[i]] < min) {
               min = fvals[itmp[i]];
               ipick = itmp[i];
            }
        }

        return ipick;
}       // end of pick parent

// note the recursive call in this method
private void tournament(int in_a_row, int nchoices, int[] picks) {
        if (in_a_row == 0) return;
        in_a_row -= 1;
        int i = rand.nextInt(nchoices);
        nchoices -= 1;
        int j = choices[i];
        choices[i] = choices[nchoices];
        choices[nchoices] = j;
        picks[in_a_row] = j;
        // recursion
        tournament(in_a_row, nchoices, picks);
}       // end of tournament

// crossover operation
private void reproduce(boolean first_child,
                        double[] child, double[][] oldpop) {
        int i, j, k, no;
        double[] pa, pb, tmp;

        if (first_child) {
           crosspt1 = rand.nextInt(nvars-1);
                           // Randomly select crossover point
           crosspt1 += 1;  // Random 1-11
           crosspt2 = rand.nextInt(nvars-crosspt1);
           crosspt2 += crosspt1;
           pa = oldpop[parent1];
           pb = oldpop[parent2];
        } else {
           pa = oldpop[parent2];
           pb = oldpop[parent1];
        }

        tmp = new double[crosspt2-crosspt1+1];

        // the following is to ensure that, after exchange, same
        // cities are not present more than once in the array
        k= 0;
        for (i=crosspt1; i<=crosspt2; i++) {        // bookkeeping
            no = 0;
            for (j=crosspt1; j<=crosspt2; j++)
                if (pa[i] != pb[j]) no += 1;
```

```
            if (no == (crosspt2-crosspt1+1)) {
               tmp[k] = pa[i];
               k += 1;
            }
        }

        for (i=0; i<crosspt1; i++) child[i] = pa[i]; // exchanging
        for (i=(crosspt2+1); i<child.length; i++) child[i] = pa[i];
        for (i=crosspt1; i<=crosspt2; i++) child[i] = pb[i];

        k = 0;
        for (i=1; i<crosspt1; i++) {                   // correcting
            for (j=crosspt1; j<=crosspt2; j++) {
                if (pa[i] == pb[j]) {
                   child[i] = tmp[k];
                   k += 1;
                }
            }
        }
        for (i=(crosspt2+1); i<child.length; i++) {
            for (j=crosspt1; j<=crosspt2; j++) {
                if (pa[i] == pb[j]) {
                   child[i] = tmp[k];
                   k += 1;
                }
            }
        }
 }      // end of method reproduce

 private void mutate(double[] child) {
        int i,j;
        double tmp;

        if (rand.nextDouble() < p_mutate) {  // Mutate this gene?
           i = rand.nextInt(nvars-1) + 1;     // random 1-11
           j = rand.nextInt(nvars-1) + 1;     // another random 1-11
           tmp = child[i];                    // swapping genes
           child[i] = child[j];
           child[j] = tmp;
        }
 }      // end of method mutate

 private void shake(double[] center, double[] x) {
        int nvars, ri;
        double tmp;

        nvars = center.length;
        while (--nvars > 0) {
              ri = rand.nextInt(nvars);       // random [0,10]
              ri += 1;                        // random [1,11]
              x[nvars] = center[ri];
              tmp = center[ri];
              center[ri] = center[nvars];
              center[nvars] = tmp;
        }
        x[0] = center[0];
 }      // end of method shake
}       // end of Genetic class
```

Listing 6.1 Genetic.java

Class TSP contains the main() method and sets up the window and menu for user interaction. Class GADialog creates a pop up dialog box (Figure 6.3) through which the user can change parameters governing the genetic algorithm. They include the number of individuals in the population, the number of generations before ending the search, the crossover probability, P_c, mutation

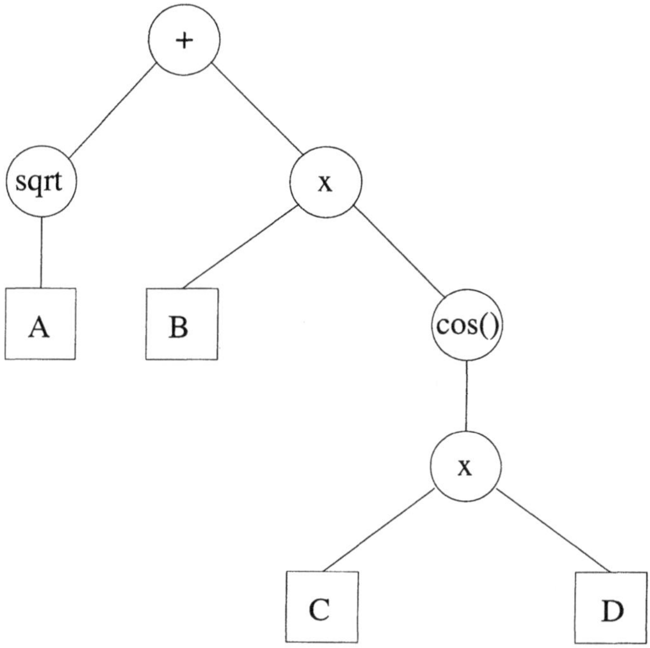

Figure 6.14. A tree example for genetic programming

probability, P_m, and the number of wins in a row for the tournament method. This proves to be helpful when we have to tune to find the set of parameters for an efficient search.

Figures 6.4 to 6.12 show screen shot of the solution after every 3 generations during the search.

We see the other equally probable solution to the problem in other runs. It is shown in Figure 6.13.

When we increase the number of cities, L, in the traveling salesman problem, the program's run time, τ, can go as a polynomial of the size of the array: $\tau \sim \alpha L^{\beta}$, where α and β are constants. Finding the shortest tour is *NP-hard* (non-deterministic polynomial-time hard); there is no short cut or smart algorithm to solve the problem quickly. Other NP-hard problems include protein design where conformations of amino acid sequences are sought to minimize the sum of interaction energies.

6.6 Genetic Programming

Consider a furnace where operating parameters are controlled and projected outputs are monitored. Dependence of the output quality on the inputs is in general described by a complicated and especially nonlinear function. We resort to engineering knowledge to model the dependence. Optimal parameters are then tuned based on the given model (function). The process is repeated until a best or satisfactory model with its optimal parameters is obtained. It would be nice to have an efficient and automatic way of obtaining the function. Genetic programming has proven to be a promising method.

In genetic algorithm, crossover and mutation operations are on fixed-length arrays that store possible solutions to the problem under study. In genetic programming, however, the operations are on the *program* itself. A program in this context is an algebraic expression such as *A*B-C/D+E*. Crossover and mutation are then performed on the expression in solving for some evaluation tasks.

The population in genetic programming now consists of expressions which are of variable length. An expression is normally represented by a tree structure as in Figure 6.14. A tree has both terminal and non-terminal components. Terminal components are usually constants or input values whilst non-terminal components are mathematical operators such as $+, -, \times, \div, \surd$, sin(), and so on. The tree of Figure 6.14 thus represents the function,

$$\text{tree} = \sqrt{A} + B \times \cos(C \times D). \tag{6.1}$$

During evolution, two winning trees are selected as parents. Crossover operations are realized by breaking up lines joining non-terminal components. Sub-trees are exchanged between the parents. Note that a tree can become indefinite long in this case and therefore it is advisable to put a limit on it. Possible mathematical operators are application specific. For instance, we may need sin() and cos() for digital signal processing. To model time series, we may define a *unit-time-delay* operator to build a memory. Therefore, before applying the technique of genetic programming, we may set aside a 'library' of operators/constants from which a tree is constructed. Mutation operation can be defined by, for example, swapping a component in the tree with that in the library.

To make things more complicated, and also more exciting, the optimal value of the constant B in Eq. (6.1) may not be known beforehands. Instead of storing a handful of constants in the library, we can tune the value of B by the method of simulated annealing, introduced in Chapter 4, at each generation during evolution. The result is a convergence toward global fitness.

In much more general cases, symbols *A, B, C*, and so on, can be instances of objects, and the mathematical operations can be replaced by the ways objects interact with one another via object methods. In this sense, genetic programming is a program that writes programs. A flurry of activities are being devoted to this area of research. Interested readers are referred to latest progresses.

6.7 Prospects

Nature exhibits not only competition for survival but also cooperation. Interdependence of different species is called symbiosis. As nature provides for us invaluable sources of inspiration, we may conceive that, some problems are easily solved by cooperation and/or co-evolution between two populations. Extension from our implementation is not difficult; cooperation might have to do with how an objective function is defined and co-evolution can be effected by changing the tournament rules, for example.

When we revisit the issue of optimal structures of neural networks (Section 5.5), we find that genetic programming may be used to help determine the optimal architecture of a neural network. Layers, neurons, and activation functions are easily represented in trees. Time delays, if present, can give rise to recurrent neural networks. Weights are adjusted by simulated annealing as mentioned above. Neural networks developed this way are called evolving neural networks. A promising application of evolving neural networks is in intelligent board-game playing programs where the esoteric board evaluation functions are developed using evolving artificial neural networks.

In the spirit of genetic programming, we might subject the cooling schedule of simulated annealing (Section 4.4) to genetic algorithm. Annealing can then become dynamic, or adaptive to a changing environment. Given various tools in this book, readers are encouraged their innovations, tackling the most challenging tasks.

6.8 Summary

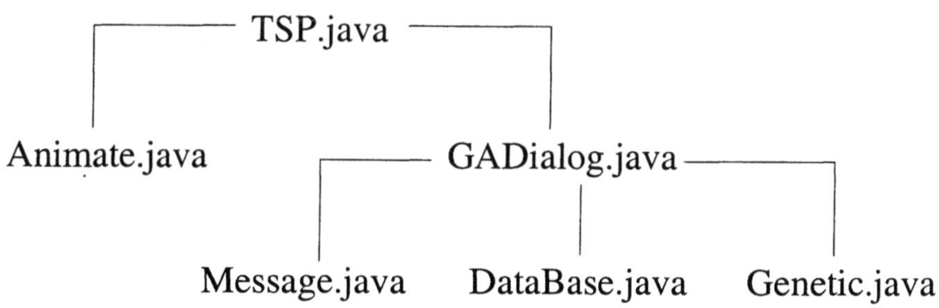

Figure 6.15. Source programs for the Traveling Salesman Problem by genetic algorithm

`Animate.java`, together with `DataBase.java`, plots the best traveling sequence in the end of each generation. `TSP.java` and `GADialog.java` are similar to `Spin.java` and `SADialog.java` of Chapter 4. They all can be obtained by minor modifications to the corresponding classes of previous chapters.

We solved a traveling salesman problem by genetic algorithm. Orders of traveling are stored in arrays which are cut in three pieces. The middle piece is swapped between parents in the crossover operation. Two elements in the array are randomly picked up and exchanged in the mutation operation. Parents who win the tournament are selected for reproduction.

Genetic programming has its genetic operations performed on structures composed of mathematic operators, constants, or even programs. Applications of genetic algorithm/programming are only limited by programmers' imaginations.

6.9 References and Further Reading

The genetic algorithm was first realized in, J. Holland, "Adaptation in Natural and Artificial Systems", University of Michigan Press, Ann Arbor (1975)

The following introduces various applications of genetic algorithm, D.E. Goldberg, "Genetic Algorithms in Search, Optimization & Machine Learning", Addison-Wesley, Reading, Massachusetts (1989)

Genetic programming was presented by, J. Koza, "Genetic Programming", *MIT Press*, Cambridge, MA (1992)

A review of evolving artificial neural networks is, X. Yao, "Evolving Artificial Neural Networks", *Proceedings of the IEEE*, **87**(9) (1999) 1423-1447

Chapter 7

MONTE CARLO SIMULATION

Monte Carlo, Monaco, hosts casinos where games of chance, such as slot machines, are played. Although a slot machine occasionally spews out chunks of tokens to lucky patrons, it, in the long run, earns a predictable fortune for the casino owner. A scientist usually devices her games of chance with various tunable parameters. After many plays, she compares the outcome with that of the real experiment. By changing the values of the parameters, she hopefully reproduces in the game the experimental result. This way, scientists gain a better knowledge of the world. In other cases where experiments are hard or expensive to carry out, simulation is the only alternative. The method of simulation or calculation that involves sampling from random numbers is coined Monte Carlo method. We show how to generate specific distributions from a uniform random distribution. We introduce and implement in the example a stochastic-volatility jump-diffusion process to simulate the complex dynamics of the price of a financial asset.

7.1 Random Number Generators

Random number generators are the hearts of any Monte Carlo methods. A uniform random number generator ideally returns, in successive calls, statistically independent numbers (`doubles`) in the half open interval [0,1). That is, u's are uniformly distributed in $0 \leq u < 1$, where u's are a sequence of numbers from the uniform random number generator. However, since computers are deterministic, true randomness is impossible. The solution is therefore to generate a long sequence of numbers that will not repeat itself until after a long cycle. The cycle of a state-of-the-art random number generator can be as large as 10^{43}. If the total number of calls to the random number generator in the simulation is less than the cycle, the finite periodicity is not a concern in practice and we call such random number generators *pseudo* random number generators.

Since a random number generator is repeatedly called during simulation, the second concern is its efficiency. A successful random number generator should be fast, saving lots of CPU time. Another consideration is its portability since the same program might be run on different platforms with different compilers by different users.

A suggestion to serious users of Monte Carlo methods is to call random number generators with caution. Adopt random number generators that are well documented and tested. Fortunately, Java comes with a robust random number generator that passes the above criteria. Successive calls to the method `nextDouble()` of an instance of the class `java.util.Random` return `doubles` that are uniformly distributed in the interval [0,1). A call to `nextInt(n)` returns an integer between `0` (inclusive) and `n` (exclusive).

7.2 Inverse Transform Method

We can calculate from first principles the average distance a photon (quantization of electromagnetic waves) travels in a material between successive interactions. The average distance is termed *mean free path*. The process is probabilistic because of the nature of quantum mechanics. The distance, x, of the photon to the next interaction is governed by the well known probability density function, $P(x)$,

$$P(x) = (1/\lambda)\exp(-x/\lambda), \tag{7.1}$$

where λ is the mean free path of the material. Different materials have different mean free paths, due to different densities, atomic numbers of the constituting elements, etc. A tissue is composed of many components characterized by their own mean free paths. Irregular geometrical shapes of the organs further complicate analytic calculation. Monte Carlo simulations are therefore routinely exercised to determine the optimal dosage to patients before radiation treatments. The question now is how to generate the probability density function from a random number generator that returns numbers uniformly distributed in [0,1).

First of all, let us construct the cumulative probability function, $C(x)$, of $P(x)$ in the following way,

$$u \equiv C(x) = \int_{-\infty}^{x} dx' P(x'). \tag{7.2}$$

Observe that $C(x)$ by such construction ranges from 0 to 1 when x runs from negative infinity to positive infinity. Equation (7.2) is equivalent to,

$$\frac{du_1}{du_2} = \frac{dC(x)/dx|x_1}{dC(x)/dx|x_2} = \frac{P(x_1)}{P(x_2)}, \tag{7.3}$$

which can be interpreted as that if we draw many random numbers in the range [0,1], the number of times they fall in du_1 divided by the number of times they fall in du_2 is equal to the ratio of the probability distribution at x_1 to that at x_2. The interpretation suggests that, x's,

$$x = C^{-1}(u), \tag{7.4}$$

are distributed according to $P(x)$ when u's are drawn from a uniform distribution in [0,1].

Taking Eq. (7.1) as an example, we have,

$$\begin{aligned} u = C(x) &= \int_{-\infty}^{x} dx'(1/\lambda)\exp(-x'/\lambda) \\ &= \int_{0}^{x} dx'(1/\lambda)\exp(-x'/\lambda) \\ &= 1 - \exp(-x/\lambda), \end{aligned} \tag{7.5}$$

which is readily inverted to give,

$$x = -\lambda\log(1-u) = -\lambda\log(u). \tag{7.6}$$

Inverse transform method is applicable when the cumulative probability function of the probability density function is known and invertible. Other such invertible examples include a Gaussian (or normal) distribution. In the chapter example, we show a different implementation of a Gaussian distribution.

7.3 Acceptance-Rejection (Von Neumann) Method

Very often the cumulative probability function is not known or the analytic form is intractable let alone its inversion. We then have to resort to the acceptance-rejection method described below.

1) find the maximum of $P(x)$ and scale it to $P'(x) = P(x)/P_{\max}$.

2) draw a random number, x, in $[a, b]$ (a and b are respectively the lower and upper bound of x).

3) draw a random number, u, in [0,1). If u is less than $P'(x)$, accept x and exit. Otherwise, go to step 2.

x's returned from the above loop distribute according to $P(x)$. In step 2, note that in order to get a random number, x, in $[a, b]$, we draw a random number, v, in [0,1) and scale it by $x = a + (b-a)v$. In the case where $b = \infty$, we can calculate $x = a[1 - \log(1-v)]$.

What the rejection method does amounts to drawing pairs of random numbers in the 2-dimensional rectangle: $(a,0)$, $(b,0)$, $(a,1)$, and $(b,1)$, and accepting the abscissa (x) whose ordinate (u) is under the curve of $P'(x)$. Obviously there are 'wastes' where the points are outside the area of $P'(x)$. Nevertheless, acceptance-rejection method can be competitive when the $C^{-1}(x)$ is very complicated. We will implement the Poisson distribution by acceptance-rejection method in the example later in the chapter. In Chapter 11, we generate a 2 dimensional distribution by the acceptance-rejection method.

7.4 Error Estimation

When reporting an experimental (or simulation) result, we show not only the value of the variable under study but also its error. The error quantifies the uncertainty of the reported value, serving as an indicator of the confidence level we have on the measurement. Errors can be classified into two categories. One is called systematic error, the other statistical error. Systematic errors come from the uncertainties/biases in the precision/accuracy of the apparatus and from environmental factors such as temperature and pressure variations during data taking. Statistical errors are of probabilistic nature: even though the same experiment with identical apparatus under the same environmental conditions is repeated, different values come out. The error from Monte Carlo simulation belongs to the latter class and is the focus of our discussion.

When we instantiate an object of the `Random` class in Java, the seed of the newly created `Random` object is set to a `long` value which is the time, measured in milliseconds, since the midnight of January 1st, 1970. We therefore normally worry little about seeding.[1] Suppose the number of trials (runs) in a Monte Carlo study is N. We implicitly assume that, in the Monte Carlos simulation, the total number of calls to the (pseudo) random number generator in the N trials does not exceed the cycle of the random number generator. If this is not sure, one way is to re-seed the random number generator before a new trial.

Suppose a Monte Carlo simulation returns an estimate for the value of the variable x. N runs of the simulation yield a distribution of the estimates for x: x_i, $i = 1, 2, \cdots, N$. The mean of the distribution, $\bar{x}$, is calculated as,

$$\bar{x} = \frac{1}{N}\sum_{i=1}^{N} x_i. \tag{7.7}$$

[1]In occasions such as debugging, however, we might want to set the seed of the random number generator to a same value in order to isolate changes due to modified program code.

The variance of the distribution, s_x^2, is,

$$s_x^2 = \frac{1}{N-1}\sum_{i=1}^{N}(x_i - \bar{x})^2 = \frac{1}{N-1}\sum_{i=1}^{N}(x_i^2 - \bar{x}^2). \tag{7.8}$$

The estimated error of the mean is,

$$s_{\bar{x}} = \sqrt{\frac{s_x^2}{N}}, \tag{7.9}$$

which is the statistical error. Finally, we report the result as,

$$x = \bar{x} \pm s_{\bar{x}}. \tag{7.10}$$

Since the statistical error, $s_{\bar{x}} \propto 1/\sqrt{N}$, to improve the estimate by a factor of 2, we need to run 4 times the number of simulations.

7.5 Multivariate Distribution with a Specified Correlation Matrix

We have so far dealt with univariate distributions. Multivariate distributions are also commonly encountered. For example, the outgoing trajectory of a particle hitting a stationary target nucleus depends on the energy and the lateral distance (the so called impact parameter) to the nucleus prior to the intersection. The distribution of the outgoing particle trajectories then depends on both the energy and impact parameter of the incoming particle. It is a bivariate distribution.

A complication in multivariate cases is that the variables might not be independent of one another. That is, a variable may *co*-vary with the other variables. The uncertainty in one variable can then propagate to the other variables. If we store the variables in a vector, the variance and covariance of the variables are then conveniently stored in a matrix called covariance matrix. An example of covariance matrix is shown in Chapter 13. The correlation matrix, C, is related to the covariance matrix, S, by the following expression,

$$S = DCD, \tag{7.11}$$

where D is a diagonal matrix whose diagonal elements are the standard deviations of the variables.

Without loss of generality, we assume the distribution of the individual variable to be Gaussian. The task is now to generate a set of Gaussian (normal) distributions with the prescribed correlation. We firstly factorize the covariance matrix into the product of an upper triangular matrix, L, and a lower triangular matrix, L^T, as below,

$$S = L^T L, \tag{7.12}$$

where L^T is the transpose of L. Next, we generate a series of N $G(0,1)$ deviates, and store them in the vector, A, of length N, N being the number of variables in the multivariate distribution and $G(0,1)$ a Gaussian distribution with mean 0 and standard deviation 1. We then form the following vector B,

$$B = AL, \tag{7.13}$$

which is a vector of Gaussian deviates with the desired covariance. The final step is to add an offset to each Gaussian deviate in order to have the desired mean. An example of a bivariate Gaussian with specified correlation is shown in the chapter example.

7.6 Stochastic-Volatility Jump-Diffusion Process

Stochastic processes are concerned with sequences of events governed in part by probabilistic laws. Namely, one of the variables in the probability distribution is time. In contrast, the angular distribution of the outgoing particle scattering off of the target nucleus should stay the same no matter when the experiment is conducted. The time invariance of physical laws is a consequence of energy conservation. In the economic setting that we are considering shortly, there appear to be few conservation laws. Part of the reason is attributed to ongoing human activities/productivities. Human brains are sources of information.

Let's look at the time series of some financial asset, such as the petroleum price, foreign exchange rate, interest rate, stock index, or electricity price. Superposed on the trend of the proportional change of the price is a random motion which can drive either up or down the value. The random motion is (almost) Gaussian distributed with a standard deviation characterizing the asset. The standard deviation is called *volatility* of the financial asset.

Volatility means risk. Speculation is one of innate human nature, which explains the omnipresence of lotteries. A large volatility could trigger larger volatilities; that is, volatilities may cluster and cascade. We therefore should consider a stochastic volatility.

A third observation in the dynamics of prices is that the arrival of a piece of 'abnormal' information can bring about a salient jump (up or down) in the asset price. Jumps occur sporadically. Examples include an outbreak of mad cow disease, causing a plummet in beef prices (because of consumers' fears of contaminations).

To simulate price movements, we then come up with the following model,

$$\frac{dS}{S} = \mu dt + s\big[dW\sqrt{1-\rho^2} + \rho dZ\big] + kdQ \tag{7.14}$$

$$d\log(s^2) = -a\big[\log(s^2) - b\big]dt + cdZ \tag{7.15}$$

where dW and dZ are independent Gaussian distributions with mean zero and standard deviation $\sqrt{dt}$. They are also called Wiener processes or Brownian motions. dQ is a Poisson distribution with rate λ, i.e., $P(dq = 1) = \lambda dt$. k is the distribution of the jump size. ρ is the correlation coefficient between the two Gaussians dW and dZ. The construction of the two correlated Gaussians follows from Eqs. (7.11), (7.12), and (7.13) with the triangular matrix L equal to,

$$L = \begin{pmatrix} c & s\rho \\ 0 & s\sqrt{1-\rho^2} \end{pmatrix} \sqrt{dt}. \tag{7.16}$$

Equation (7.14) states that the proportional increment in the asset price diffuses along the deterministic trend of $dS/S = \mu dt$. The diffusion is modeled by a Gaussian distribution with mean zero and variance $s^2 dt$ (for $\rho = 0$ for simplicity). Jumps further contribute to the proportional change in the asset price. The average number of jumps per unit time is set to λ of the Poisson distribution. Once a jump occurs, its size is sampled from the jump size distribution, k. Equation (7.15) describes the dynamics of the volatility. When there is no diffusion, i.e., when $c = 0$, $\log(s^2)$ goes like,

$$\log\left(s(t)^2\right) = b + \left(\log\left(s(0)^2\right) - b\right) \exp(-at). \tag{7.17}$$

The feature is that the volatility is believed to eventually revert to the mean after turmoils; a is the reverting rate and b gives the long-term volatility. c contributes to volatility of the volatility movement.

Equations (7.14) and (7.15) are general enough to model the dynamics of most financial assets. Variants exist, for example, to model the term structure of interest rates. A caveat in model building: take the parsimonious approach. Try first the model that has the least parameters. Considering the above jump-diffusion model, it is nontrivial to separate, in data, the jump from the diffusion contribution. Estimating the size distribution and frequency of the jumps is also difficult, especially when the amount of data is limited. In addition, volatility is an elusive quantity since it is not a traded asset. Calibrating Eqs. (7.14) and (7.15) is not straightforward. Econometricians work their hardest to unfold the statistics.

7.7 A Cash Flow Example

We demonstrate an application of Monte Carlo methods in this section. Suppose I commute between Vancouver and New York monthly. The air fare fluctuates, which, among other factors, might be mirroring cost of the fuel. I use Eqs. (7.14) and (7.15) to model the price movements which could tell me the amount of money I need to put aside now for the known itineraries in one year. The assumption is that I don't book in advance and therefore I don't lock in the cost now. Similar and probably more realistic situations occur to a company.

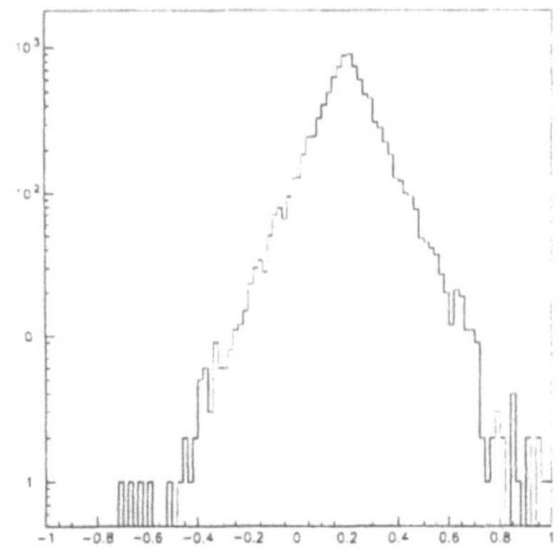

Figure 7.1. An exponential-decay jump-size distribution. The ordinate is in log scale.

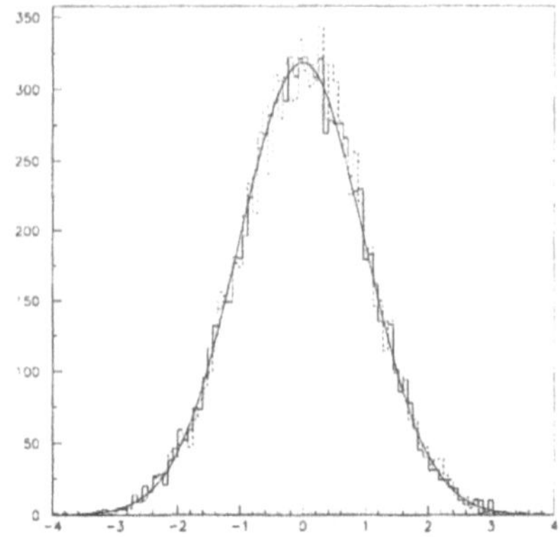

Figure 7.2. Two independent unit Gaussian deviates

For example, a company expects to receive a steady flow of cash in foreign currency in the coming year. The amount that will be received in domestic currency is however uncertain due to fluctuations in the exchange rate. Operational cost of a company is better below revenues. Estimation of the present value of the cash flow can be done using Monte Carlo techniques. Another example is an insurance company which models various claims to make sure that requirements of the financial well-being of the insurance company is met.

Before generating possible price movements by Monte Carlo, we delineate a jump size distribution,

$$k = \begin{cases} (1/\tau)\exp\big((-x+\mu_j)/\tau\big) & \text{if } x > \mu_j \\ (1/\tau)\exp\big((x-\mu_j)/\tau\big) & \text{if } x < \mu_j \end{cases} \tag{7.18}$$

which is, when $\mu_j = 0$, an exponential decay function that is symmetrical about the vertical axis. A positive μ_j simply shifts the whole distribution rightward, giving more upward jumps than downward jumps. τ quantifies the width of the distribution. A histogram of such a jump size distribution is shown in Figure 7.1. The number of samplings is 10,000.

We write down the corresponding difference equations of the stochastic-volatility jump-diffusion process to assist in coding,

$$\begin{aligned} \log\Big(\frac{S_{i+1}}{S_i}\Big) &= \mu/\Delta + s_{i+1}\big[W_{i+1}\sqrt{1-\rho^2} + \rho Z_{i+1}\big]/\sqrt{\Delta} \\ &\quad + \begin{cases} 0 & \text{if } n = 0 \\ k_1 + k_2 + \cdots + k_n & \text{if } n > 0 \end{cases} \end{aligned} \tag{7.19}$$

$$\log\Big(\frac{s_{i+1}^2}{s_i^2}\Big) = a\big[b - \log(s_i^2)\big]/\Delta + cZ_{i+1}/\sqrt{\Delta} \tag{7.20}$$

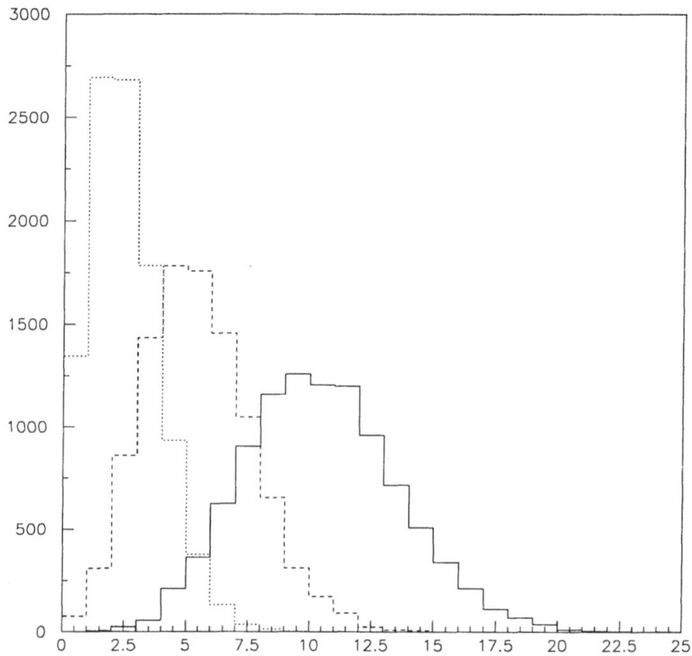

Figure 7.3. Histograms of Poisson distribution. Solid, dashed, and dotted lines are for Poisson of mean (i.e. λ/Δ) equal to 2, 5, and 10, respectively.

where Q is a Poisson counter: $P(Q = n) = \exp(-\lambda/\Delta)(\lambda/\Delta)^n/n!$ with n a non-negative integer. If we let dt to be one day, the rates in the model are then daily rates and Δ is the number of observations per day. A Java implementation of Eqs. (7.18), (7.19) and (7.20) is given in Listing 7.1. Successive calls to the method `gaussian()` in the `MonteCarlo` class give two independent series of values. Each series is Gaussian distributed with mean 0 and standard deviation 1. An example of the two independent unit Gaussian deviates is shown in Figure 7.2. The number of draws is 10,000. The solid curve in the figure is a fit of the counts to a Gaussian function. Dashed histogram is the other Gaussian. Figure 7.3 shows the histograms of returns from successive calls to the `poisson()` method in the `MonteCarlo` class. Figure 7.4 is a dialog box interface for the user to input initial values of the asset and daily volatility, and to change the parameter values of the stochastic-volatility jump-diffusion

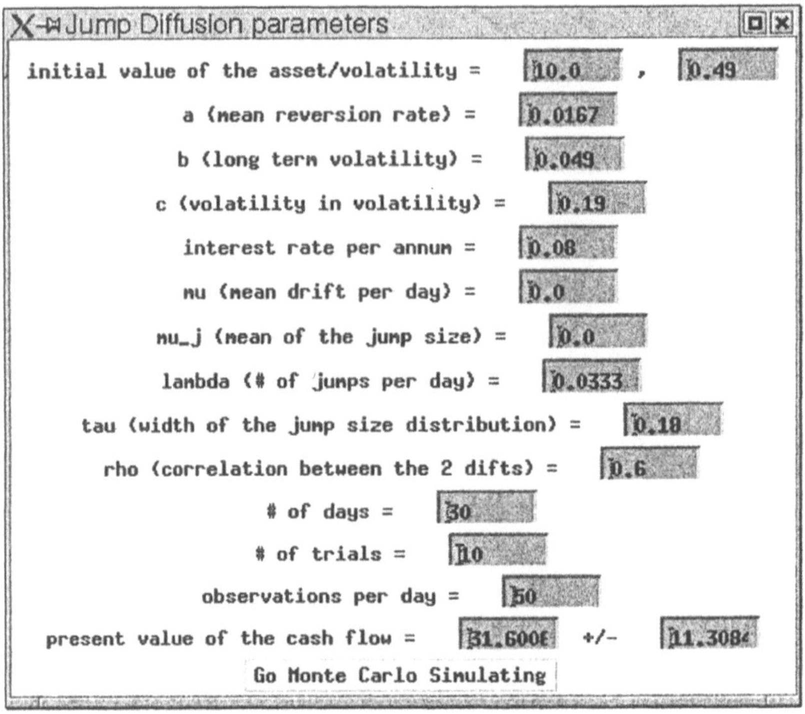

Figure 7.4. Graphical user interface for the stochastic-volatility jump-diffusion process

process. Figures 7.5 to 7.10 show possible price movements given different values of $\mu, a, b, c, \lambda, \tau, \mu_j, \rho$ and so on. Red (green) vertical lines in the figures indicate upward (downward) jumps.

```
/*      Sun-Chong Wang
        TRIUMF
        4004 Wesbrook Mall
        Vancouver, V6T 2A3
        Canada
        e-mail: wangsc@triumf.ca

        MonteCarlo.java generates paths by sampling from
        the stochastic-volatility jump-diffusion model,
        Eqs. (7.18), (7.19), (7.20) */

import java.lang.*;
import java.util.*;
import java.util.Random;

public class MonteCarlo {
 Random rand;
 int nSteps,delta;
 static int nTrials;
 static double S_0=10.0,v_0=0.05;
 static double a,b,c,mu,mu_j,tau,lambda,r;
 double payOff,payOff_error,rho,g1,g2;
 double[] path;
```

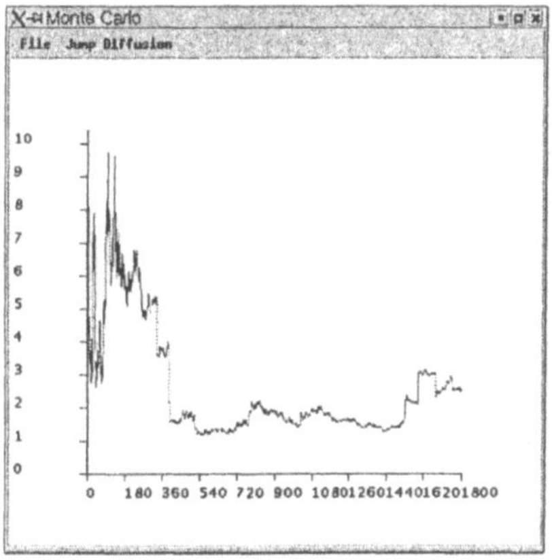

Figure 7.5. A reversion to a smaller long-term volatility

Figure 7.6. A reversion to a larger long-term volatility

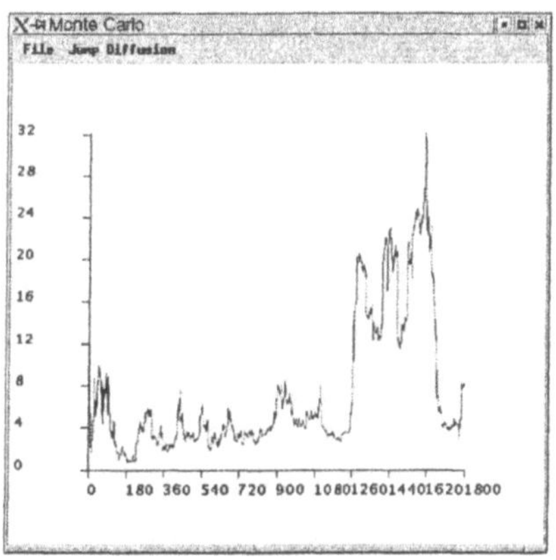

Figure 7.7. A higher jump frequency

Figure 7.8. Larger jump sizes

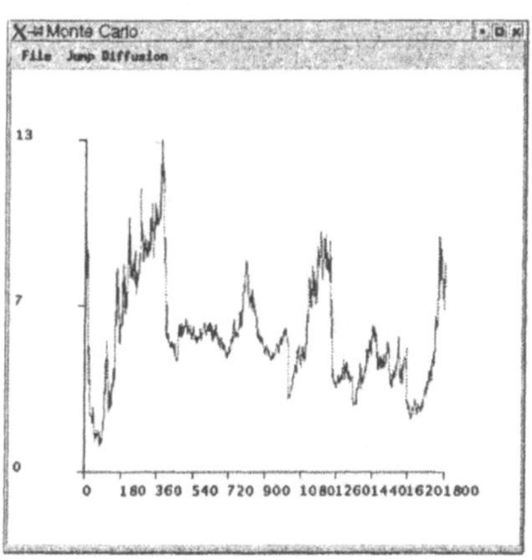

Figure 7.9. Mostly downward jumps

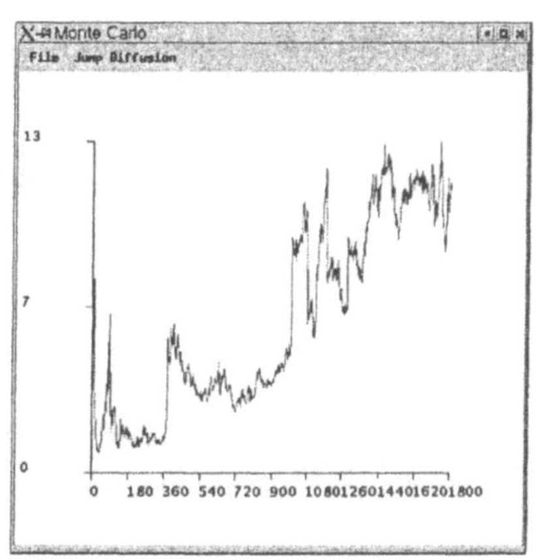

Figure 7.10. A positive mean return

```
int[] jumps;

MCDialog parent;

public MonteCarlo(MCDialog parent) {
        this.parent = parent;
        rand     = new Random();
        a        = 0.0;
        b        = 0.1;
        c        = 0.3;
        r        = 0.08;
        mu       = 0.0;
        mu_j     = 0.2;
        tau      = 0.1;
        lambda   = 0.1;
        delta    = 5;
        rho      = 0.0;
        nSteps   = 1000;
        nTrials  = 100;
        payOff_error = 0.0;
        payOff   = 0.0;
        System.out.println("random number = " + rand.nextDouble());
}       // end of MonteCarlo class constructor

public void go() {
        Initialize();
        Mean();
}

private double exponential(double tau) {
// generates distribution (1/tau)exp(-x/tau)
        double rnd;

        do {
           rnd = rand.nextDouble();
        } while (rnd == 0.0);
        return -tau*Math.log(rnd);
}

private double jump_size(double mean, double tau) {
        double tmp;

        tmp = exponential(tau);
        if (rand.nextDouble() > 0.5) return tmp+mean;
        else return -tmp+mean;
}

private void gaussian() { // Gaussian distribution
        double v1,v2,r2,tmp;

        do {
           v1 = 2.0*rand.nextDouble() - 1.0;
           v2 = 2.0*rand.nextDouble() - 1.0;
           r2 = v1*v1 + v2*v2;
        } while (r2 >= 1.0 || r2 == 0.0);

        tmp = Math.sqrt(-2.0*Math.log(r2)/r2);
        g1 = v1*tmp;    // two independent unit normals
        g2 = v2*tmp;    // g1 and g2
}

private int poisson(double lambda) { // Poisson distribution
        int n=0,k=1;
        double tmp,A=1.0,oldA=A;

        do {
           A = rand.nextDouble()*oldA;
           tmp = Math.exp(-lambda);
```

```
            if (A < tmp) n = k - 1;
            else k += 1;
            oldA = A;
        } while (A > tmp || A == tmp) ;

        return n;
}

public void Initialize() {
        path = new double[nSteps];
        jumps = new int[nSteps];
}

private void generatePath() {
        int J;
        double logS,v,logS_next,v_next,g;

        logS = Math.log(S_0);
        v = v_0;

        for (int i=0; i<nSteps; i++) {
            gaussian();                // get 2 independent normals
            v = 2.0*Math.log(v);
            // stochastic volatility, Eq. (7.20)
            v_next = v + a*(b-v)/delta + c*g2/Math.sqrt(delta);
            v_next = Math.exp(v_next/2.0);
            g = Math.sqrt(1.0-rho*rho)*g1 + rho*g2;
            // diffusion, Eq. (7.19)
            logS_next = logS + mu/delta + v_next*g/Math.sqrt(delta);
            J = poisson(lambda/delta);     // now the jumps
            if (i > 0) jumps[i-1] = J;
            for (int j=0; j<J; j++) logS_next += jump_size(mu_j,tau);
            path[i] = Math.exp(logS_next);
            v = v_next;
            logS = logS_next;
        }
}

public void Mean() {
        int month;
        double tmp,payOff2;

        payOff = 0.0;
        payOff2 = 0.0;
        for (int i=0; i<nTrials; i++) {
            generatePath();
            tmp = 0.0;
            for (int j=0; j<nSteps; j++) {
                // 30 days per month
                month = (j+1)/(delta*30);
                if ((j+1)%(30*delta) == 0) {
                    // discounting to get the present value
                    tmp += path[j]*Math.exp(-r/12.0*month);
                }
            }
            payOff += tmp;
            payOff2 += tmp*tmp;
        }

        payOff /= nTrials;                       // mean, Eq. (7.7)
        payOff_error = Math.sqrt((payOff2/nTrials- // Eq. (7.9)
                                  payOff*payOff)/nTrials);
        // plotting
        parent.parent.db.pathToDraw(path,jumps);
}       // end of Mean
```

```
}        // end of MonteCarlo class
```

Listing 7.1 MonteCarlo.java

7.8 Variance Reduction Techniques

Equation (7.9) reads that to reduce the variance of an estimate by a factor of 2, we need to double the number, N, of simulation runs. In some cases, we can double N without doubling the number of random number generator calls in the program. Imagine that an estimation results from a number ϵ drawn from a Gaussian random distribution of mean zero and standard deviation one. Since ϵ is symmetric about zero, it is equally likely that a $-\epsilon$ is drawn in a separate run. We can therefore perform two estimates, one using ϵ, the other $-\epsilon$, whenever an ϵ is generated. The estimate using $-\epsilon$ is called *antithetic variable* and the technique of combining the two estimates antithetic variate method. Also, since $1-u$ is randomly distributed between 0 and 1 if u is randomly distributed between 0 and 1, estimates with $1-u$ are antitheses of estimates with u. Note that antithetic variate method is easily implemented when the specified distribution is generated by *single* calls to the random number generator as in the inverse transform method.

The second technique we are introducing is the *control variate* method. We use Monte Carlo method to get an unbiased estimate, $\bar{X} = (\sum_{i=1}^{N} X_i)/N$, of X, when we do not have a closed form solution to X. Suppose there exits Y which is similar to X and that we know its exact solution to be Y_0. At the same time of calculating X_i, we calculate Y_i and evaluate the quantity $X_i + c(Y_i - Y_0)$, where c is a constant. If $\bar{Y}$ is an unbiased estimate of Y, $\bar{X} + c(\bar{Y} - Y_0)$ is also an unbiased estimate of X. Now, in the equality,

$$Var\big(X + c(Y - Y_0)\big) = Var(X) + c^2 Var(Y) + 2c\,Cov(X,Y), \quad (7.21)$$

the minimum of $Var\big(X + c(Y - Y_0)\big)$ occurs when c is,

$$c = -\frac{Cov(X,Y)}{Var(Y)}. \quad (7.22)$$

Suppose, for example, X is positively correlated with Y; an overestimate X_i is accompanied by an overestimate Y_i. The overestimate is however offset by $c(Y_i - Y_0)$ in $X_i + c(Y_i - Y_0)$ with the choice of c in Eq. (7.22). Y is called a control variate for X. Unfortunately, $Cov(X,Y)$ and $Var(Y)$ are usually not known beforehand and need to be determined from simulation.

7.9 Summary

With simple modifications to corresponding classes in previous chapters, we can get `JumpDiffusion.java`, `Plotter.java`, `DataBase.java`, and `MCDialog.java` for the stochastic-volatility jump-diffusion application.

We implemented methods that return exponentially, normally, and Poisson distributed values from the uniform random number generator provided in Java's utility package. Combining these distributions allowed us to model such complex processes as the evolution of the price a financial asset. Monte Carlo methods are used in simulations by engineers, modeling by astronomers, and

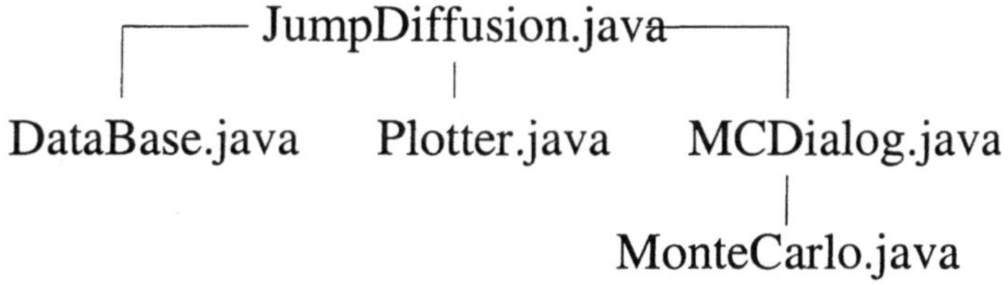

Figure 7.11. Source programs for the stochastic-volatility jump-diffusion process

in risk assessment by managers. We showed how to generate correlated multivariate Gaussian distributions. We introduced two variance reduction techniques. Different seeds are set to represent different realizations of the same underlying process. Statistical uncertainties should be calculated and quoted together with the mean in the end of the simulation. A uniform random number generator that passes randomness tests and is well documented is the one to work with.

7.10 References and Further Reading

Textbooks on Monte Carlo simulation are, R.Y. Rubinstein, "Simulation and the Monte Carlo Method", John Wiley and Sons, Inc., New York (1981), and L. Devroye, "Non-Uniform Random Variate Generation", Springer-Verlag, New York (1986)

Textbooks on probability and statistics are, G. Cowan, "Statistical Data Analysis", Oxford University Press, Oxford (1998), and R.J. Barlow, "Statistics: a Guide to the Use of Statistical Methods in the Physical Sciences", John Wiley, New York (1989)

Chapter 8

MOLECULAR DYNAMICS

Molecular dynamics simulation is widely used in, for example, molecular biology, material engineering, and surface physics to study protein folding, structure defect, and crack propagation. Structures of proteins, working parts of a cell, are believed to determine their functions, the knowledge of which helps understand life and also accelerate drug design. In this chapter, we establish the connection between microscopic motions of atoms and their macroscopic properties. A molecular dynamics example is then provided to simulate release of particles from a compartment (vaporization of a droplet).

8.1 Computer Experiment

A molecular dynamics simulation can be thought of as an experiment performed on a computer. Computer experiments have an equivocal role in scientific research. They are not real experiments, nor are they pure theories. It is nevertheless an economical and sometimes the only feasible way of investigation. Collisions of galaxies and cosmological evolution by molecular dynamics, *in silico* biological experiments, and reaction of fuel in a combustion engine are among the examples.

In molecular dynamics simulation, an atom (or molecule) interacts with the other atoms (molecules) in the system. The interaction is modeled by a *potential* which is a function of positions of the atoms. The spatial gradient of the potential, of the atom of interest, gives the force on the atom. The formalism is a result of total energy conservation. Newton's second law in classical mechanics relates the force on the atom to its acceleration,

$$\frac{F_i}{m_i} = -\frac{1}{m_i}\nabla_i V(r_1, r_2, \cdots, r_N) = a_i = \frac{\partial v_i}{\partial t} = \frac{\partial^2 r_i}{\partial^2 t}, \tag{8.1}$$

where F_i is the force on atom i, m_i, r_i, v_i, a_i are respectively the mass, position, velocity, and acceleration of atom i. $V(r_1, r_2, \cdots, r_N)$ is the potential energy of the system and N is the total number of the atoms in the system. Once we have the acceleration of the atom, its position at the next instance of time can be obtained by integrating over time. This procedure is repeated and we get the evolution of the system. Note that, unlike Monte Carlo simulations, molecular dynamics is deterministic: given the same initial conditions, the system evolves the same way and gives the same result.

Before we link the microscopic law of motion to the macroscopic behavior of the system, we need to justify the legitimacy of Eq. (8.1), i.e. the classical Newton's equations of motion, in molecular dynamics simulation. Molecules are microscopic entities and shouldn't they be reined by quantum mechanical laws? Associated with every atom is a quantity called de Broglie thermal wavelength, Λ, expressed as,

$$\Lambda = \text{constant}\frac{1}{\sqrt{MT}}, \tag{8.2}$$

where M and T are respectively the mass of the atom and temperature of the system in kilogram and Kelvin. If the thermal wavelength is much shorter than the characteristic length, a, of the system (average separation of atoms in this case), 'particle' nature of the atom dominates and motion of the atom can be described by classical mechanics. When the thermal wavelength approaches the characteristic length, 'wave' nature of the atom begins picking up and we need to add quantum corrections to the equation of motion (semi-classical mechanics). When $\Lambda \gtrsim a$, quantum mechanics reins and the atom is described by its wavefunction.[1] From Eq. (8.2), we note that for light elements, such as H_2, He, and Ne at low temperatures, Eq. (8.1) may not be valid. Fortunately, for most of the systems (solids, liquids, and gases) we are interested at normal conditions, we can safely integrate Newton's second law of motion, Eq. (8.1).

8.2 Statistical Mechanics

A cup of coffee can contain 10^{24} molecules, each molecule, under mutual interaction, moving toward its own destination. To describe the coffee at an instance of time, we might have needed 6×10^{24} numbers: 3 position coordinates plus 3 velocities in each spatial dimension for each molecule. The 6×10^{24} numbers define a *configuration* of the system. What we, human beings in a macroscopic world, care about is the temperature of the coffee; we do not need details of the coordinates and velocities of every molecule. Statistical mechan-

[1] A quantum mechanical formalism by the method of Feynman's path integral is introduced in Chapter 10.

ics has established formula relating the microscopic states (configurations) to the macroscopic property (temperature) of a system in equilibrium. [An equilibrium state means a state far away from (long after) external disturbance.]

Consider now two cups of coffee. If they differ in their initial conditions, the configurations developed later in the mugs are different. They nonetheless can have the same temperature! Configurations are different, but they have the same weight of testifying things. To be fair, therefore, statistical mechanics calculates a macroscopic quantity (temperature) by averaging over *all possible* different configurations. In this way, to get the temperature, you can think of measuring the temperature of many (infinite, in principle) cups of coffee at the same time, and the average of them is the temperature you get. This is not very economical in practice, but it is exactly what statistical mechanics tells us. Those many cups of coffee form an *ensemble* and the averaging over them is called ensemble average.

Macroscopic quantities of interest include kinetic energy, total energy, temperature, pressure, pair correlation function, and so on. The last quantity is interesting in that it signifies whether a system is in a solid, liquid, or gaseous state.

8.3 Ergodicity

If however the cup of coffee can be well isolated after it is prepared, we can measure the temperature of it as many times as we want. This average should also give us the temperature of the coffee. This is indeed true and supported by the *ergodic hypothesis* of statistical mechanics,

$$\textbf{ensemble average} = \textbf{time average}. \tag{8.3}$$

Configurations of the coffee at different instances of time are equivalent to different configurations of the coffee prepared at the same time. After an indefinite long time, the configuration of the single cup should have passed through all possible configurations, according to the ergodic hypothesis.

In molecular dynamics simulation, once the system reaches an equilibrium state, we start calculating and accumulating quantities for macroscopic states such as temperature, pressure as we visit successive configurations while integrating, step by step, Newton's equations of motion, Eq. (8.1). After the prescribed number of time steps is reached, we find the average values to get the macroscopic quantities of interest. Ergodic hypothesis is assumed in molecular dynamics simulation.

8.4 Lennard-Jones Potential

The inter-atomic potential is the key machinery we need in a molecular dynamics simulation. In fact, molecular dynamics is often used in the other way around to find out the unknown interaction form. Here we introduce an inter-

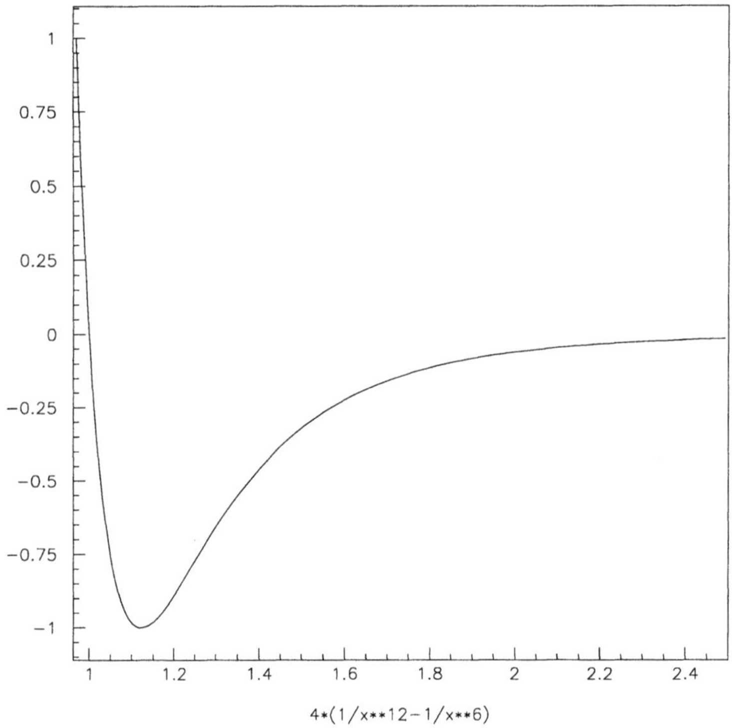

Figure 8.1. The Lennard-Jones potential

atomic potential which describes very well the interaction between noble atoms or neutral molecules, i.e. the Lennard-Jones potential (Figure 8.1),

$$V_{LJ}(r) = 4\epsilon\left[\left(\frac{\sigma}{r}\right)^{12} - \left(\frac{\sigma}{r}\right)^{6}\right], \tag{8.4}$$

where r is the distance between the two atoms, ϵ scales the magnitude of the potential energy and σ is the location where the potential energy is zero. It is known that atoms/molecules exhibit weak attraction to each other. The attraction is long ranged and is expressed by the second term on the right of Eq. (8.4) with the correct r dependence due to dipole-dipole interaction. The $1/r^6$ term dominates when r is large. However, if two atoms are brought closer, the interaction becomes repulsive, from the exclusion of the atomic electrons. The repulsion is modeled by the first term of Eq. (8.4) which is dominant at short distances. The total potential energy of the system is the sum over pairs of atoms,

$$V = V(r_{12}) + V(r_{13}) + \cdots + V(r_{23}) + \cdots = \sum_{i} \sum_{j>i} V(r_{ij}). \tag{8.5}$$

Energy and distance in molecular dynamics are very small numbers in units of the standard SI system (Systeme International d'Unites). It is therefore customary in molecular dynamics to express energy in units of ϵ, length in units of σ and mass in units of atomic mass. For argons in this reduced unit system,

$$\begin{aligned}
\text{time is } \sigma\sqrt{\text{mass}/\epsilon} &= 3.4 \times 10^{-10} m \times \\
&\quad \sqrt{6.69 \times 10^{-26} kg / 1.65 \times 10^{-21} J} \\
&= 2.17 \times 10^{-12} s, \\
\text{velocity is } \sqrt{\epsilon/\text{mass}} &= 1.57 \times 10^{2} m/s, \\
\text{force is } \epsilon/\sigma &= 4.85 \times 10^{-12} N, \\
\text{pressure (for a 2 - d system) is } \epsilon/\sigma^2 &= 1.43 \times 10^{-2} N/m, \\
\text{and temperature is } \epsilon/k_B &= 120\, K, \\
\text{where } k_B &= 1.38 \times 10^{-23} J/K
\end{aligned} \tag{8.6}$$

is Boltzmann's constant.

8.5 Velocity Verlet Algorithm

Once we have the potential, we calculate the gradient to get the force, which immediately gives the acceleration. The next step is to integrate the acceleration to give the position of the atom at the next time step, according to Eq. (8.1). There are a couple of time integration algorithms popular among researchers. They differ in the precision and memory requirement. Since hardware memories are nowadays relatively cheap, we introduce the so-called *velocity Verlet* algorithm which compromises no precision.

The algorithm is basically a Taylor series expansion of the variable,

$$r(t + \Delta t) = r(t) + v(t)\Delta t + \frac{1}{2} a(t) \Delta t^2, \tag{8.7}$$

which updates positions of the atoms. Once we have the new positions, new forces and thus accelerations are available via the potential energy, Eq. (8.5), which is a function of solely positions. To be able to keep iterating Eq. (8.7), we need to update velocities, which is accomplished by,

$$v(t + \Delta t) = v(t) + \frac{1}{2} \big[a(t) + a(t + \Delta t) \big] \Delta t. \tag{8.8}$$

The truncation error of the above velocity Verlet approximation is of the order of $(\Delta t)^3$. When we choose Δt to be 0.01 in the reduced unit, after a large number of iterations, errors still accumulate and the venerable energy/momentum conservation can be lost. In practice in the simulation, we therefore from time to time calculate and rebalance total energy and momentum.

8.6 Correcting for Finite Size and Finite Time

Very often we are simulating bulk properties of a material. The number of atoms in present molecular dynamics simulations is less than 1 million which is far less than the number of atoms of the read world system under study. In order to reduce the surface effect in such a small system (box), we normally impose periodic boundary conditions where atoms leaving the boundary re-enter the system from the opposite face of the box. A one dimensional box with such a periodic boundary condition is a circle; A two dimensional box is a torus (or doughnut, bagel), etc.

The maximum number of atoms in the simulation is certainly limited by the computing capacity. We do not know a priori if, say, 1 million atoms is biasing the simulation result. Or should we go to 10 million? On the other hand, perhaps 100,000 atoms is adequate. Similar question arises to the size, L, of the box. One way to check it out is to perform simulations of the quantity of interest, A, at different box size L and compare the results with the following relation,

$$A(L) = A_{L\to\infty} + \frac{b}{L^c}, \tag{8.9}$$

where $A_{L\to\infty}, b, c$ are fitting parameters and $A_{L\to\infty}$, the value of A when L goes to infinity, is considered the 'true' value of the variable A. Equation (8.9) is supposed to take care of the so-called *finite size effect.*

The time average method, replacing the ensemble average via the ergodic hypothesis, is used to calculate equilibrium quantities. We have said nothing about how to start the simulation. The easiest way is to initialize both position and velocity vectors to random values. The system is supposed to settle down (relax) to an equilibrium soon. We then start summing $A(t)$ as t goes on. To speed up relaxation, we can also start outright from a legitimate configuration by sampling positions and/or velocities from a known distribution corresponding to a macroscopic state. An example is the Maxwell-Boltzmann distribution which gives the probability distribution of velocities for a system at a specific temperature. Either way, to ensure that the system has relaxed into equilibrium and that the the number of time iterations is large enough, we poll the value of A at different time and plot the results against the following relation,

$$A(t) = A_{t\to\infty} + \beta \exp(-t/\gamma), \tag{8.10}$$

where again $A_{t\to\infty}, \beta, \gamma$ are fitting parameters and $A_{t\to\infty}$, the value of A as the simulation time t gets indefinite long, is the equilibrium value of the variable.

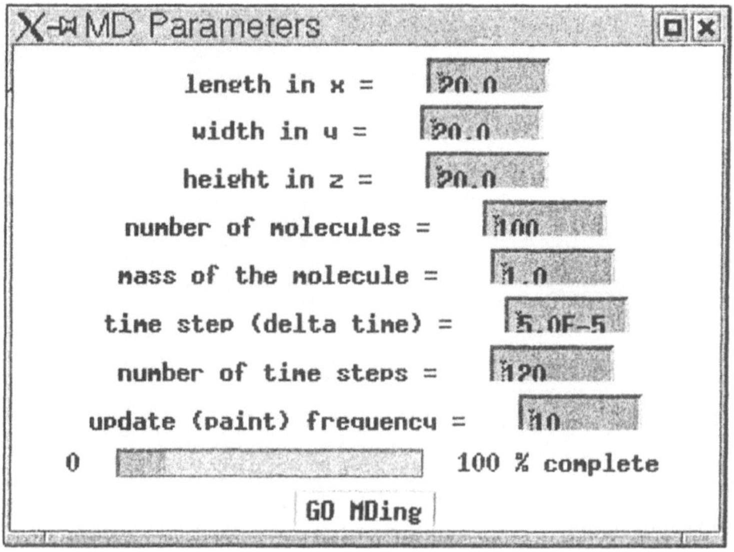

Figure 8.2. User interface for the molecular dynamics simulation

8.7 An Evaporation Example

Listing 8.1 (`MD.java`) gives the code for a 3 dimensional molecular dynamics simulation implementing the Lennard-Jones potential of Eq. (8.4) and velocity Verlet algorithm of Eqs. (8.7) and (8.8). Figure 8.2 shows the interface dialog box for users to input simulation parameters such as dimensional sizes of the box, number of atoms in the box, size of a time step, number of time steps, and the frequency of graphing the atoms on screen. In the animation from Figure 8.3 to 8.14, configurations of the system are plotted every 10 time steps. In the beginning of the simulation, 100 atoms are jammed in a small box at the center of the box. Atoms are then let go. Subsequent loci of the individual atoms are determined by the potential, Eq. (8.5), and Newton's equations of motion, Eq. (8.1). Since our system is 3 dimensional, x and y coordinates of the atom are used to locate the atom represented on the Java canvas as a solid red circle. The z coordinate is then used to determined the radius of the circle; the closer the atom to the user, the bigger the radius (cf. Section 4.5).

```
/*      Sun-Chong Wang
        TRIUMF
        4004 Wesbrook Mall
        Vancouver, V6T 2A3
        Canada
        e-mail: wangsc@triumf.ca

        MD.java implements Lennard-Jones potential and velocity
        Verlet approximation for molecular dynamics simulation */

import java.lang.*;
```

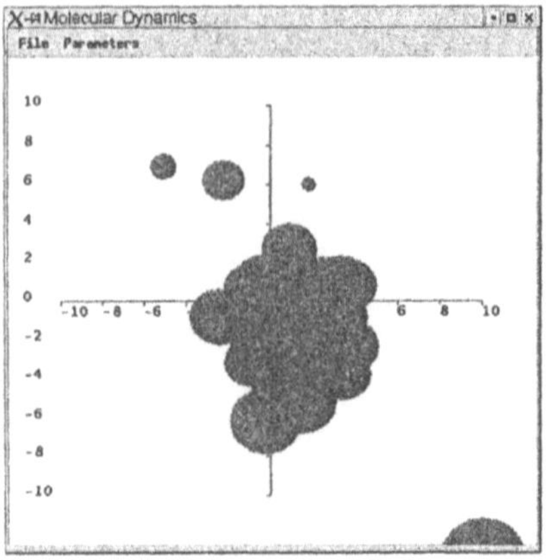

Figure 8.3. animation of the molecular dynamics simulation begun

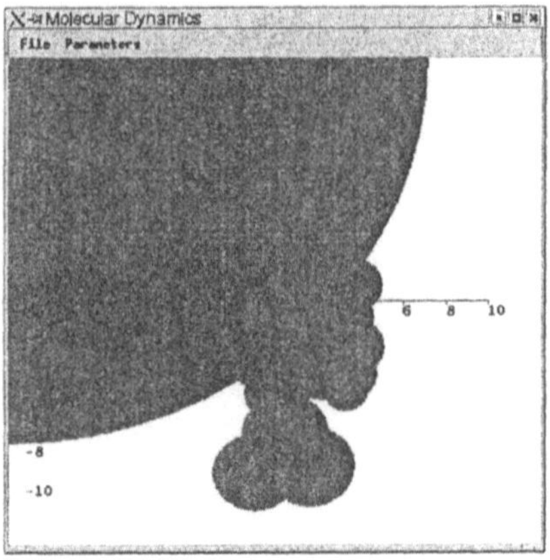

Figure 8.4. animation continued

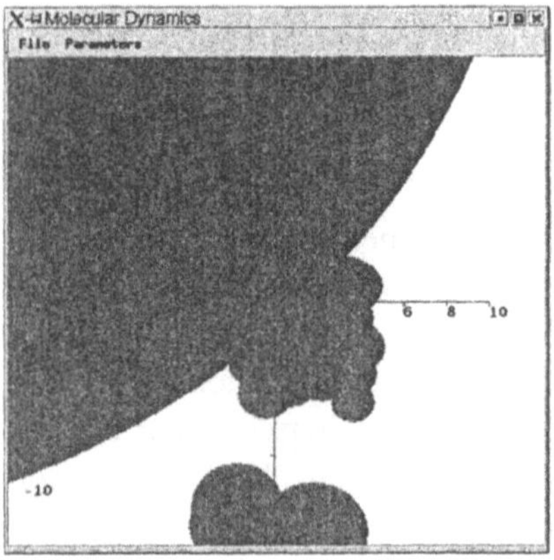

Figure 8.5. Animation continued

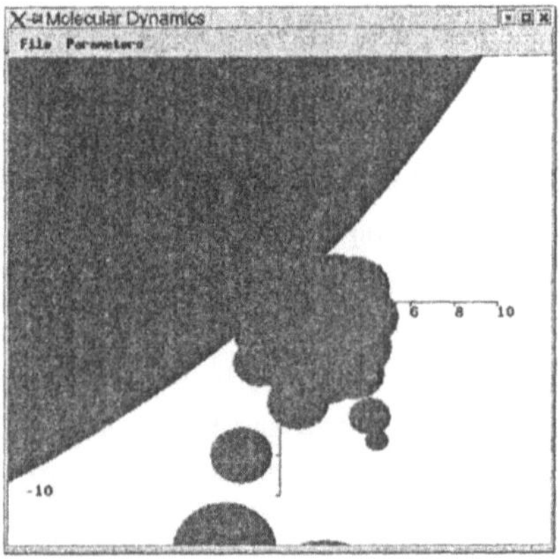

Figure 8.6. Animation continued

```
import java.util.*;
import java.util.Random;

public class MD implements Runnable {
 Random rand;

 final int D = 3;  // dimension of the box
 int iTime, ifreq;
 int num_molecules;
 double V,dt,dt2,mass,one_over_mass;
 double[][] r,r_next,v,v_next,a,a_next,dVdr;
 double[][] plot_r;
 double[] L;

 MDDialog parent;
```

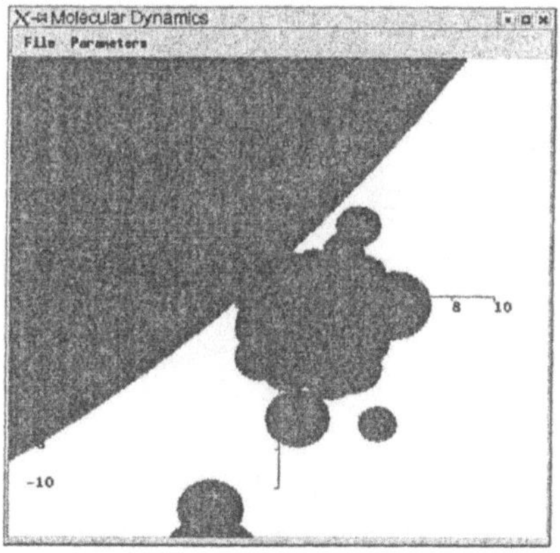

Figure 8.7. Animation continued

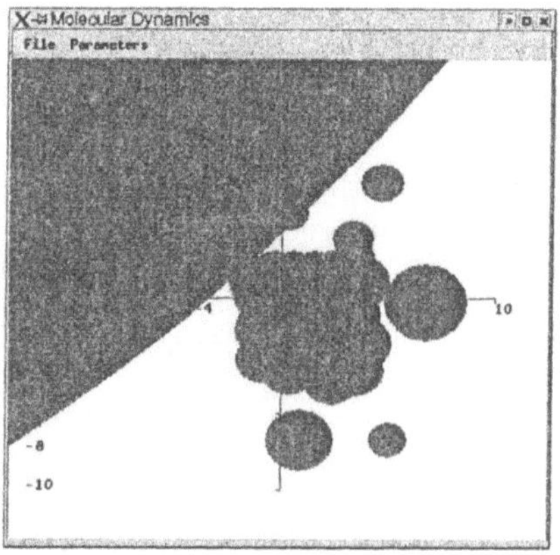

Figure 8.8. Animation continued

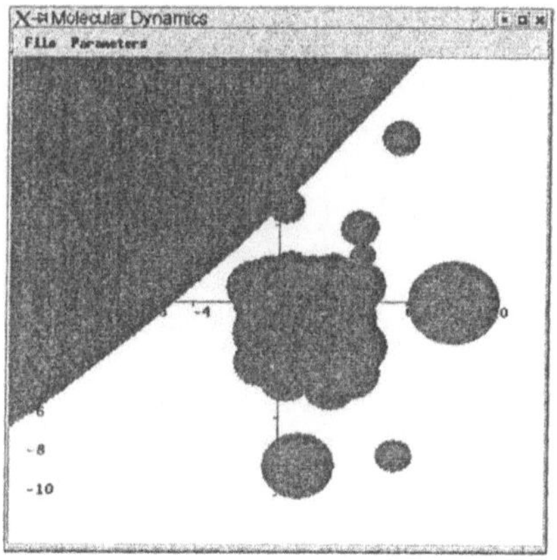

Figure 8.9. Animation continued

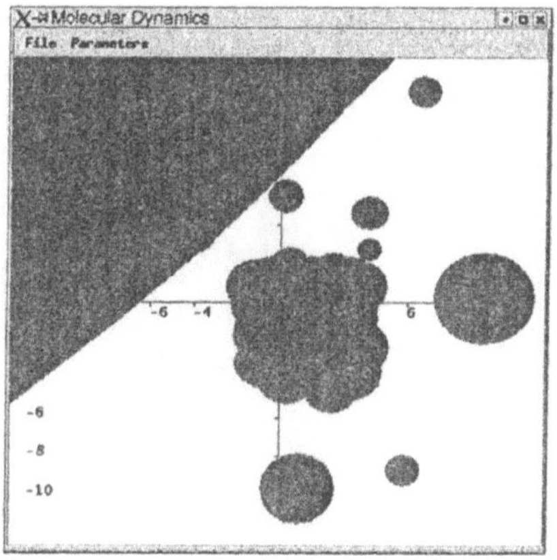

Figure 8.10. Animation continued

```
public MD(MDDialog parent) {
        this.parent = parent;

        rand = new Random(100L);
        System.out.println("random number = " + rand.nextFloat());

        num_molecules = 0;
        mass = 1.0;
        one_over_mass = 1.0/mass;
        dt = 0.01;
        dt2 = dt*dt;
        L = new double[D];
        iTime = 10000;
        ifreq = 100;
}       // end of MD class constructor

public void run() { // this method runs in a thread
```

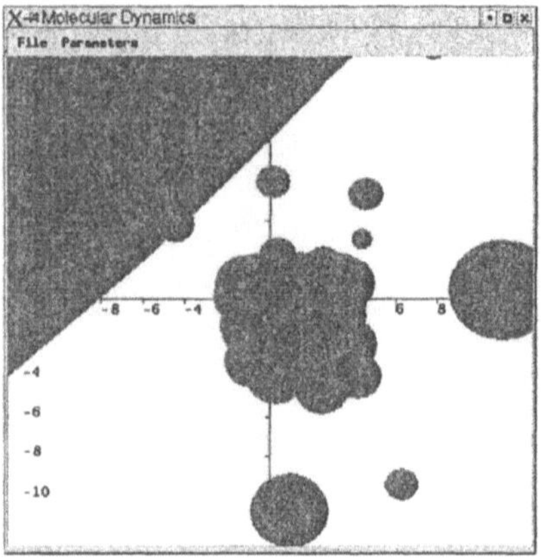

Figure 8.11. Animation continued

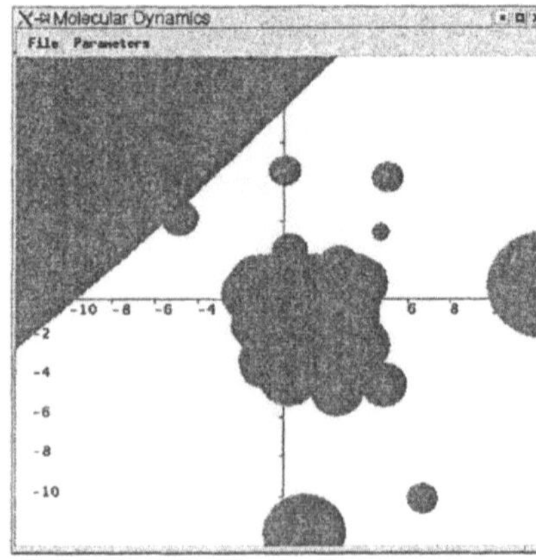

Figure 8.12. Animation continued

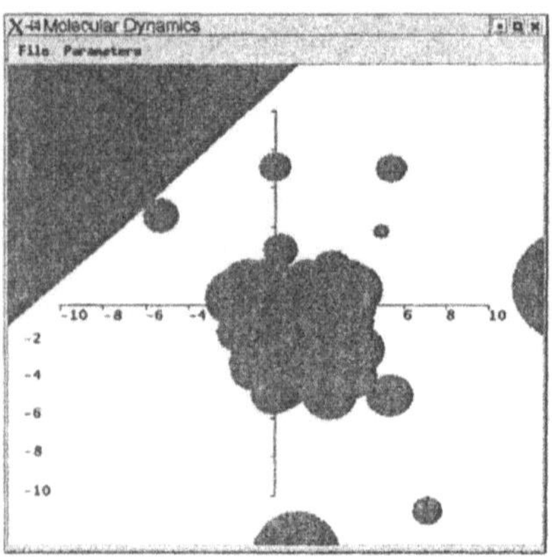

Figure 8.13. Animation continued

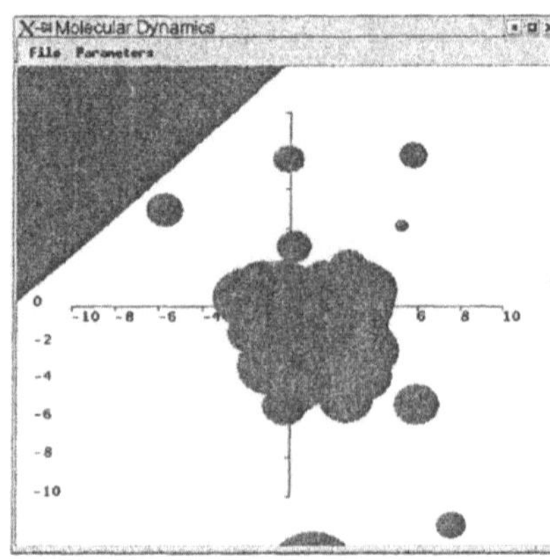

Figure 8.14. Animation finished

```
for (int i=0; i<iTime; i++) {
    VelocityVerlet();
    if (((i+1)%ifreq) == 0) {
       for (int j=0; j<r.length; j++)
           for (int k=0; k<r[0].length; k++)
               plot_r[j][k] = r[j][k];
       parent.parent.db.DrawMolecules(plot_r,L);
       try {
           parent.thread.sleep(10000); // for screen shooting
       } catch (Exception errr) {
           System.out.println("Errorr : "+errr.getMessage());
       }
       System.out.println("iteration = "+i);
    }
    parent.jbar.setValue(Math.round((i+1)/((float)iTime)
```

```
                                                        *100.0f));
        }
 }

 public void Setup(int num_molecules, double mass, double dt,
                   double XBound, double YBound, double ZBound) {
        if (this.num_molecules != num_molecules) {
           this.num_molecules = num_molecules;
           r = new double[num_molecules][D];
           v = new double[num_molecules][D];
           a = new double[num_molecules][D];
           r_next = new double[num_molecules][D];
           v_next = new double[num_molecules][D];
           a_next = new double[num_molecules][D];
           dVdr   = new double[num_molecules][D];
           plot_r = new double[num_molecules][D];
        }
        L[0] = XBound;
        L[1] = YBound;
        L[2] = ZBound;
        this.dt = dt;
        dt2 = dt*dt;
        this.mass = mass;
        one_over_mass = 1.0/mass;
        // prepare for the initial configuration:
        // molecules are jammed in the central core
        // note that molecule coordinates are within -L/2 and +L/2
        rand.setSeed(100L);
        for (int i=0; i<r.length; i++)
            for (int j=0; j<r[0].length; j++)
                r[i][j] = (2.0*rand.nextDouble()-1.0)*L[j]/2.0/4.0;
        // v's start from zero, but we can get a's
        // from the initial configuration
        dLJdr(r);
        for (int i=0; i<r.length; i++)
            for (int j=0; j<r[0].length; j++)
                a[i][j] = -one_over_mass*dVdr[i][j];
 }      // end of Setup

 private double Periodic(double r, double b) {
         // periodic boundary conditions
         while (r < -b/2.0) {
            r += b;
         }
         while (r > b/2.0) {
            r -= b;
         }
         return r;
 }

 private void VelocityVerlet() {                  // Eqs. (8.7), (8.8)
        for (int i=0; i<r.length; i++) {
            for (int j=0; j<r[0].length; j++) {
                r_next[i][j] = r[i][j]+v[i][j]*dt+0.5*a[i][j]*dt2;
                // check boundary conditions
// uncomment below to have periodic boundary conditions
//                r_next[i][j] = Periodic(r_next[i][j],L[j]);
            }
        }

        dLJdr(r_next);   // calculate the potential gradient

        for (int i=0; i<r.length; i++) {
            for (int j=0; j<r[0].length; j++) {
                a_next[i][j] = -one_over_mass*dVdr[i][j];
                v_next[i][j] = v[i][j] + 0.5*(a[i][j]+a_next[i][j])*dt;
```

```
                }
            }

            for (int i=0; i<r.length; i++) {
                for (int j=0; j<r[0].length; j++) {
                    r[i][j] = r_next[i][j];
                    v[i][j] = v_next[i][j];
                    a[i][j] = a_next[i][j];
                }
            }
    }       // end of VelocityVerlet

    private void dLJdr(double[][] r) {
            // Lennard-Jones potential. results are stored in dVdr[][]
            double V,d2,d6;

            V = 0.0;
            for (int i=0; i<r.length; i++)
                for (int k=0; k<r[0].length; k++)
                    dVdr[i][k] = 0.0;

            for (int i=0; i<(r.length-1); i++) {
                for (int j=(i+1); j<r.length; j++) {
                    d2 = 0.0; // squared distance of the pair of molecules
                    for (int k=0; k<r[0].length; k++) {
                        d2 += (r[i][k]-r[j][k])*
                              (r[i][k]-r[j][k]);
                    }
                    d6 = d2*d2*d2;
                    V += (1.0/d6/d6-1.0/d6); // LJ potential
                    for (int k=0; k<r[0].length; k++) {
                        dVdr[i][k] += (-12.0/d6/d6/d2+6.0/d6/d2)*r[i][k];
                        dVdr[j][k] += (-dVdr[i][k]);
                    }
                }
            }

            V *= 4.0;
            for (int i=0; i<r.length; i++)
                for (int k=0; k<r[0].length; k++)
                    dVdr[i][k] *= 4.0;
    }
}           // end of MD class
```

Listing 8.1 MD.java

8.8 Summary

Again, `Evaporation.java`, `DataBase.java`, `Plotter.java`, and `MDDialog.java` can be readily obtained by using corresponding classes in previous chapters as templates.

We animated a toy evaporation process by molecular dynamics simulation with the velocity Verlet algorithm. Appropriate parameters have to be instituted for a more serious and realistic simulation. We addressed the issue of finite size/time effect and their corrections. Time averaging for macroscopic quantities in molecular dynamics simulation is adequate as long as the statistical mechanics,ergodic hypothesis is justified. Macroscopic properties of interest are related to microscopic motions of the constituent particles in statistical mechanics.

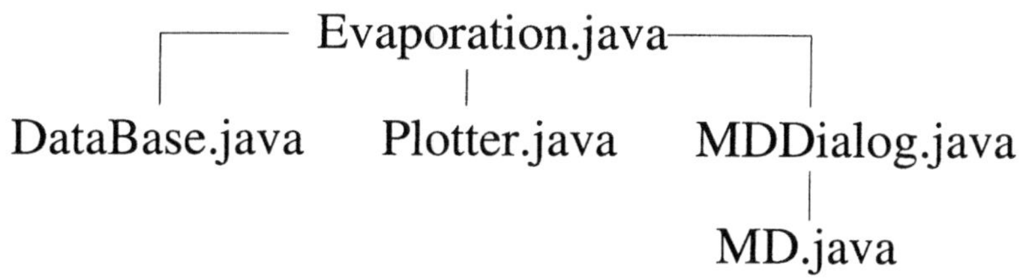

Figure 8.15. Source programs for the molecular dynamics simulation

8.9 References and Further Reading

Two textbooks on molecular dynamics are, J.M. Haile, "Molecular Dynamics Simulations: Elementary Methods", Wiley, New York (1992), and D.C. Rapaport, "The Art of Molecular Dynamics Simulation", Cambridge University Press, Cambridge, England (1995)

Textbooks on statistical mechanics include, S.-K. Ma, "Statistical Mechanics", World Scientific, Singapore (1985), and L.E. Reichl, "A Modern Course in Statistical Physics", University of Texas Press, Austin (1980)

Chapter 9

CELLULAR AUTOMATA

An ant, compared with other species, is a simple creature. Yet, a colony of ants forms a single complex hierarchical system, which, in some sense, can be more efficient than other ingregarious, yet more advanced, organisms. Ants utilize simple protocols in communications between each other. An ordered system is thereby formed and individuals know where/how to efficiently locate/transport foods. Careful examinations revealed that the system works from bottom up. The result is remarkable: the whole is greater than the sum of the parts.

An ant is considered a cellular automaton. Imagine that in the Internet is distributed a web of simple 'computer agents', among which is set up a proper set of reaction rules. The effectiveness of the web of agents could be astounding; they can be designed to effectively search for or filter information, for instance. Cellular automata simulation can also be used to help design exits and signs for an efficient evacuation of a stampeding crowd in a theater.

9.1 Complexity

Although there are phenomena that exhibit chaotic behaviors, triggered by a minute difference in the initial conditions, there is evidence that in biology there exists anti-chaos which, together with natural selection, shapes the way species evolve. Many systems in nature look complex and irregular on one scale. But the irregularities look similar on all scales. For example, coastlines/riverlines look similar whether they are viewed from an airplane or satellite; galaxies and clusters of galaxies look similar; changes in stock prices between different time intervals look very similar.

Interactions of cells, with interconnecting signal transmitting molecules, form a complex network that defines the living system. An example is the immune system which responds to foreign agents in the organism. At any different instance of time during the life cycle of an organism, different patterns of genetic activity in the genome (the complete set of genes of an organism) manifest to regulate different functions of the cells/organs. A genome is a self-regulatory complex system.

There are, conservatively, a dozen definitions of complexity. Here we are concerned about the emergence of self-similarity and/or self-regularity on the verge of chaos. Many of these properties have been modeled by cellular automata.

9.2 Self-Organized Criticality

Consider hand clapping of audience in the end of a performance. Immediately after the show, the stage darkens and the curtain falls. The audience remains intoxicated with the great performance. Then a few hands start clapping, applauding the great moment the performers just presented. As more and more hands join the applause, a pattern of clapping emerges, and suddenly all the hands in the audience clap at a climax rate. Amateur shows by, for example, the street artists in New York, might not be as warmly received, so the extent of hand clapping varies. Self-organized waving has also been experienced in a sports stadium.

In earth science, researchers analyze earthquakes of varying magnitudes and poise what conditions trigger an earthquake. The distribution of the number of earthquakes N of magnitude k versus k follows a power law,

$$N(k) \propto \frac{1}{k^\alpha}, \tag{9.1}$$

where α is some positive number characterizing the phenomenon.[1] The above are examples of what is called *self-organized criticality*. Self-organizing refers to the fact that the system is closed with excitations coming from within, such as the audience in the show or the tectonic structure beneath the earthquake-prone region. Criticality means spread of the effect to all scales.

9.3 Simulation by Cellular Automata

The phenomena of complexity or, in a stricter sense, self-organized criticality, in social, economical, biological, ecological, or physical science are vastly studied by the method of cellular automata on computers. In a cellular automata simulation, space is discretized into lattice sites (nodes), and time is discretized into short steps. The state of a lattice site is also discrete. It can be used to represent, for example, how many sand 'particles' are present at a site in the study of avalanches in a sand pile. New state of the site is determined

[1] In the World Wide Web, the number of webpages, N, having k hyperlinks into (out of) a page also follows a power law. This is called *scale free*. Examples of scale-free network include power line grids, traffic networks, networks of neural cells, networks of metabolic reactions, networks of scientific citations, etc.

by its present state and the current state of the neighboring sites. For example, with simple rules adding more sands to randomly selected sites, sites are toppled whose number of sands exceeds a certain threshold. In cellular automata, states of the sites of the system are updated in every time step.

Cellular automata method is therefore capable of simulating dynamics (or evolution) of a system. We can find application of cellular automata in a constellation of different fields. In the following, we demonstrate its successful application to hydrodynamics.

9.4 Lattice Gas Automata

Navier-Stokes differential equations are usually the starting point for fluid dynamics. A typical way of exploring is to transform the differential equation into difference equation which is then integrated from given initial values. However, because of rounding and truncation, processes that are necessary to store real valued numbers in finite bit-width computer memories, errors are introduced and gradually accumulated. They can become huge, rendering the calculation unacceptable. Although some tricky (and often awkward) patches to the difference equation can come to the rescue, a totally different approach can be more satisfying.

In the lattice gas automata approach, an underlying lattice is employed to define the spatial coordinates. For simplicity, we consider a 2-dimensional triangular lattice, where every lattice point has six nearest neighbors (Figure 9.1). Next, there can be zero or up to six gas 'molecules' at each lattice point. The molecule moves at one speed and heads for one of the six directions: north-west, west, south-west, south-east, east, and north-east, as in Figure 9.1. There is however an exclusion rule that no two (or more) molecules can have the same direction of motion at any site.

The state of a lattice point can be conveniently represented by a `byte` (8-bit in Java), where bit j of one (or zero) stands for the presence (or absence) of a molecule moving in direction j at the time. Note that more bits can be introduced to represent if the molecule is molecule A or molecule B when two fluids are simulated.

After introducing the representation, we need to specify the updating rules, which is the other essential ingredient of cellular automata. As mentioned, the state of a site at the next time step depends on its state and the states of its neighboring sites at present time. Updating rules for the lattice gas automata are simply to preserve mass (particle number) and momentum. It is exactly these fundamental rules, together with the Boolean nature of the representation, that make lattice gas automata stable compared with the difference equation method. Figure 9.2 depicts the conservation rules. With the byte representation of a state, the rules can be translated into a table. Updating of the gas, in the program, then amounts to looking up the table. The coding is there-

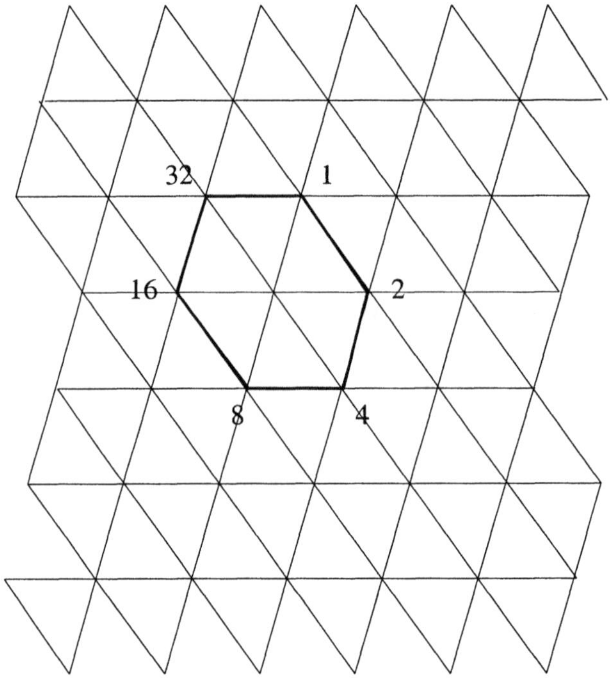

Figure 9.1. The triangular grid representing a surface

fore very straightforward. See, however, the next section for elaboration and subtleties due to geometry.

It has been shown that 2-dimensional cellular automata on triangular lattice reproduce most of the important aspects of real 2-dimensional gases. Navier-Stokes equations can in fact be reproduced from the automata with the mass and momentum conserving rules. For 3-dimensional gases, however, a face-centered hyper-cubic (FCHC) lattice has to be constructed, serving as the underlying lattice, in order for the system to have the desired isotropy.

The collision rules that were specified limit the viscosity of the fluid that can be attained in the simulation. Variants of the model exist. For example, more than one value of speed can be used to increase the Reynolds number of the system (which is proportional to the velocity and density of the gas system). Although other methods of computational fluid dynamics offer higher precision, lattice gas automata are suited to simulate systems where correlations play an important role such as in reaction-diffusion problems. Lattice gas automata also find a niche in studying complex fluids whose Navier-Stokes equations are not (well) known.

(a)
(b)
(c)
(d)
(e)

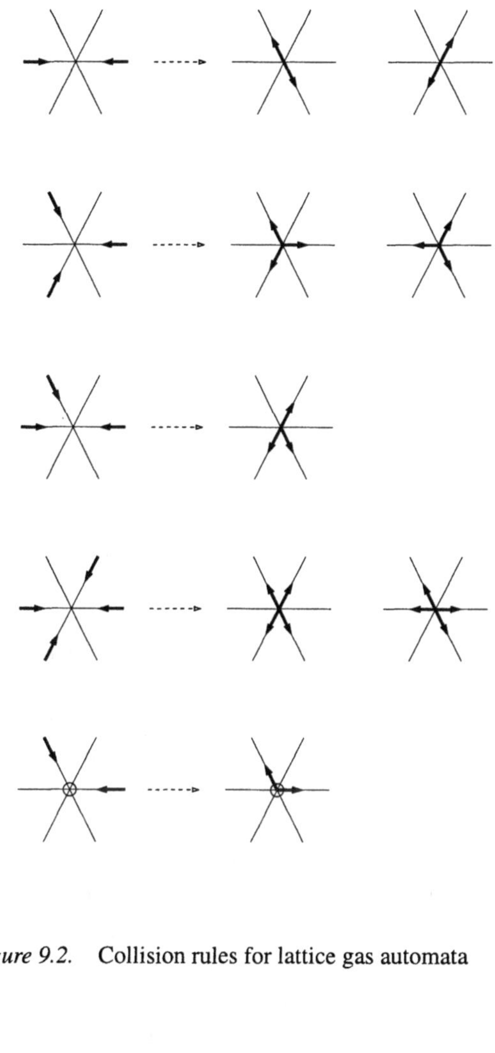

Figure 9.2. Collision rules for lattice gas automata

9.5 A Hydrodynamic Example

The example demonstrates a lattice gas automata in two dimensions. Gas molecules travel along the lines in Figure 9.1 and collide with one another at the nodes. The six velocities are labeled in the way shown in Figure 9.1. They are encoded by the first 6 bit of a `byte` variable. For example, a value of 36 (=00100100) at node i represents that two molecules appear at node i, one traveling at direction southeast, the other northwest,

Each time step of the simulation consists of two operations: collision and transportation, which are performed system-wise. The collision rules of Figure 9.2 respect both mass and momentum conservation. To code them, consider,

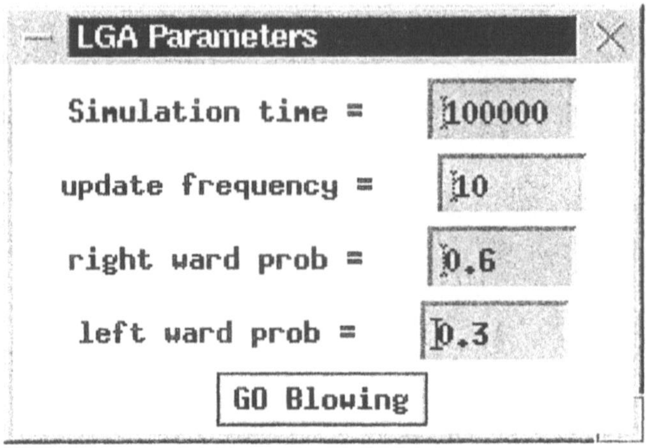

Figure 9.3. The dialog box which reads lattice gas automata inputs

for example, the rule of Figure 9.2 (a). In the above byte representation, it reads 18 $\rightarrow$ 9 or 36.

A boundary node can be represented by the 7th bit of its value. So, for example, before the collision step, a value of 72 (=01001000) at node i represents a particle, coming from the northeast, reaches the boundary site at i. We impose the so-called *no-slip* boundary condition: molecules hitting the boundary reverse their directions of motion. Therefore, for instance, 72 (=64+8) before collision is turned to 65 (=64+1) after collision, and the like. One example is depicted in Figure 9.2 (e).

Furthermore, we can impose the periodic boundary condition in the horizontal direction: molecules reaching the right (east) edge and are still moving eastbound are translated to the left edge of the space. Initial molecules are prepared and injected into the space from the left with user defined eastbound and westbound probabilities. Note that in the transportation step, even rows and odd rows are transported differently (cf. Figure 9.1).

The system in our simulation consists of 400$\times$400 nodes. To get physical quantities, we form coarse grained values by averaging over domains. There are, say, 10 by 10 nodes in one domain. The total number of molecules in the domain is summed and then divided by the number of nodes in the domain. Similar averages are obtained for domain velocities, which are plotted on the screen along the course of the simulation. Note that in the program, we made extensive use of bitwise operations. Listing 9.1 implements all the methods for the lattice gas automaton simulation. Figure 9.3 shows the dialog box for user input.

```
/*      Sun-Chong Wang
```

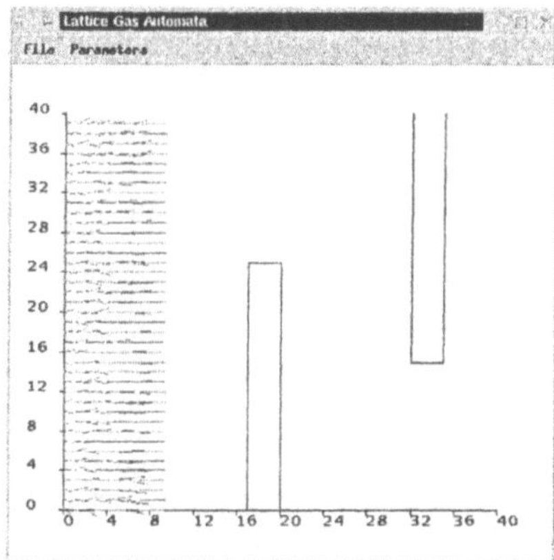

Figure 9.4. animation of the lattice gas automata begun

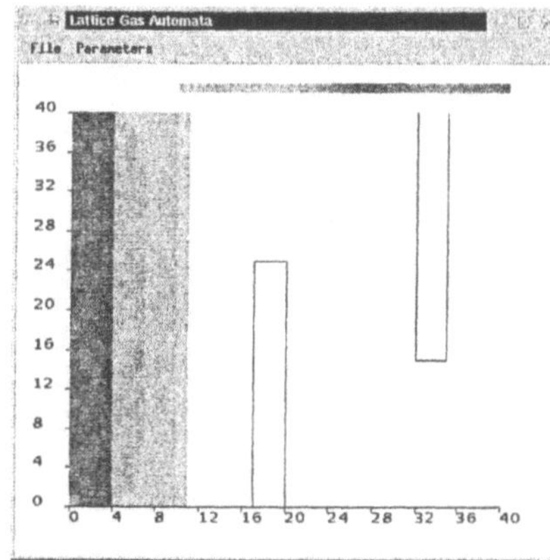

Figure 9.5. animation continued (mass density)

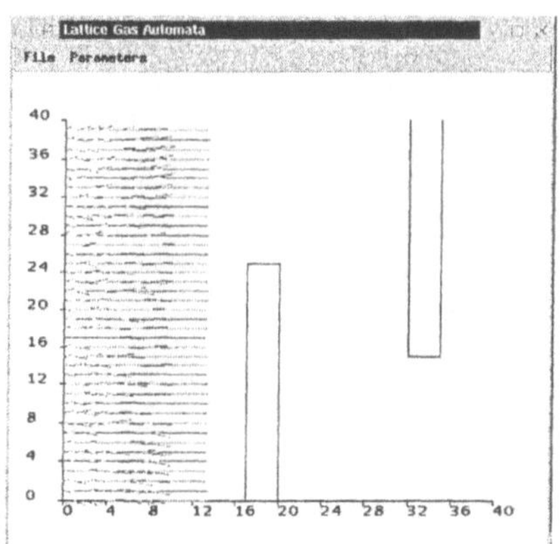

Figure 9.6. animation continued (velocity field)

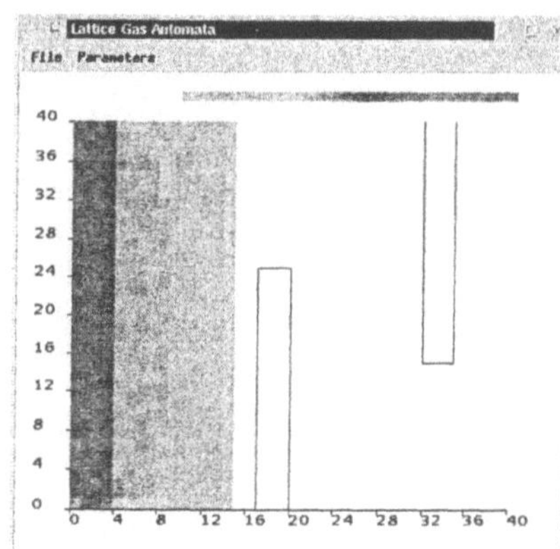

Figure 9.7. animation continued (mass density)

```
        TRIUMF
        4004 Wesbrook Mall
        Vancouver, V6T 2A3
        Canada
        e-mail: wangsc@triumf.ca

        LGA.java updates and transports the gas by the method
        of lattice gas automata */

import java.lang.*;
import java.util.*;
import java.util.Random;

public class LGA implements Runnable {
 Random rand;
```

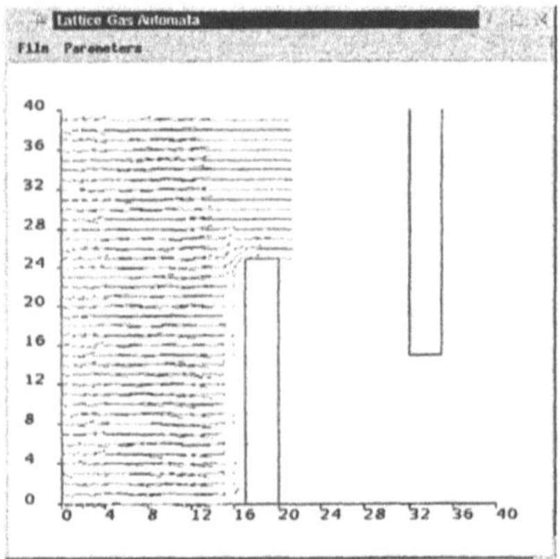

Figure 9.8. animation continued (velocity field)

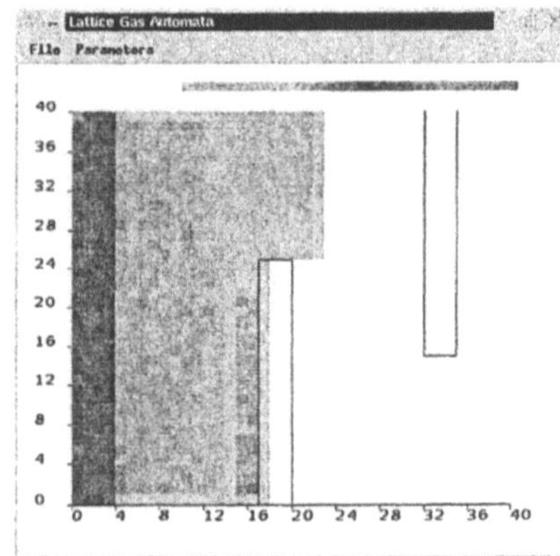

Figure 9.9. animation continued (mass density)

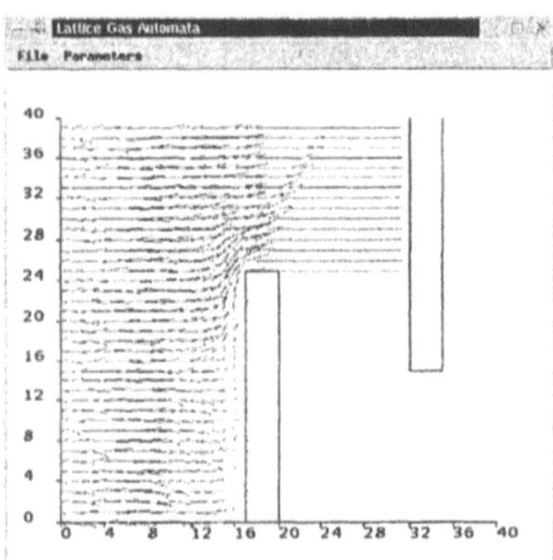

Figure 9.10. animation continued (velocity field)

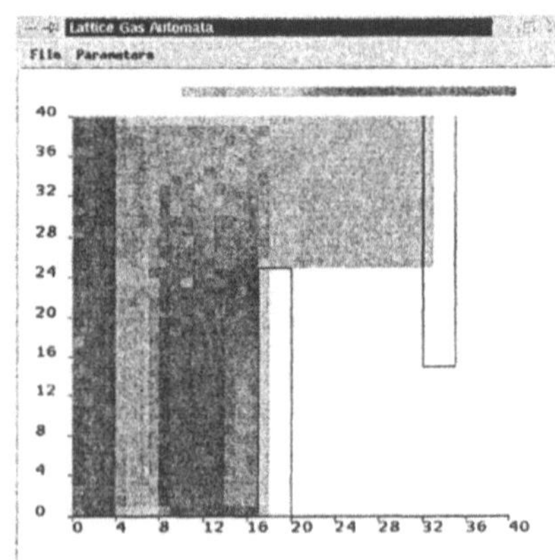

Figure 9.11. animation continued (mass density)

```
int XSize, YSize, Domx, Domy;
int iTime, ifreq;
double[][] mass;
int current_m, Init_m;
double p_left, p_right;
double[][] Vx, Vy;        // domain averaged velocities
byte[][] state, state_tmp;

LGADialog parent;

public LGA(LGADialog parent) {
        this.parent = parent;

        rand     = new Random();
```

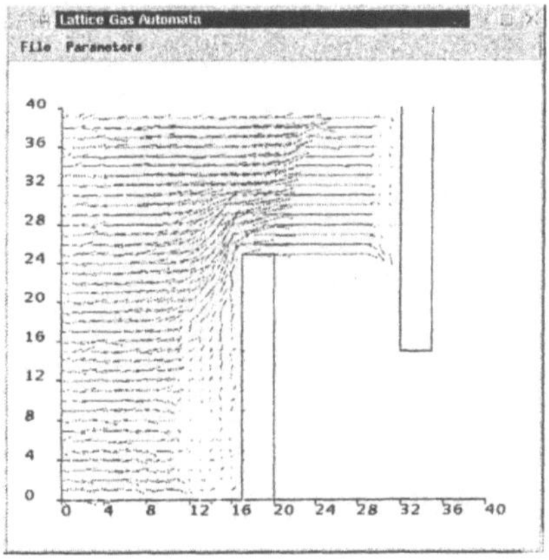

Figure 9.12. animation continued (velocity field)

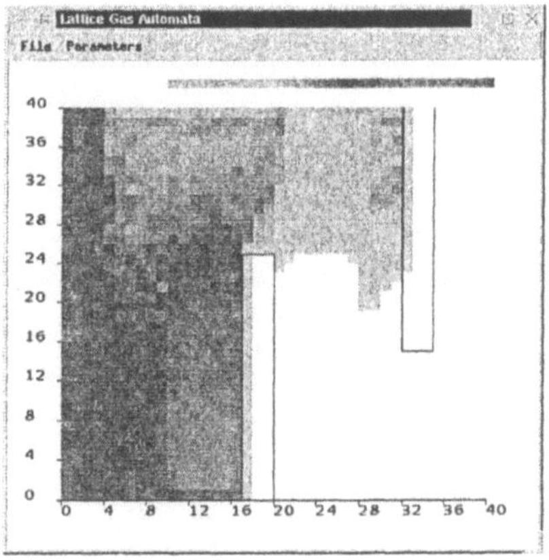

Figure 9.13. animation continued (mass density)

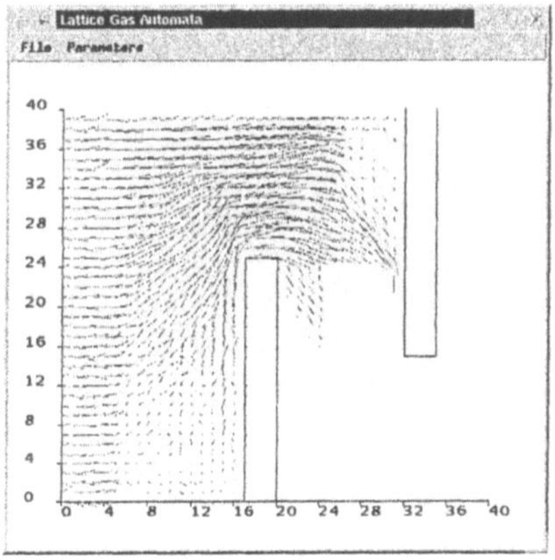

Figure 9.14. animation continued (velocity field)

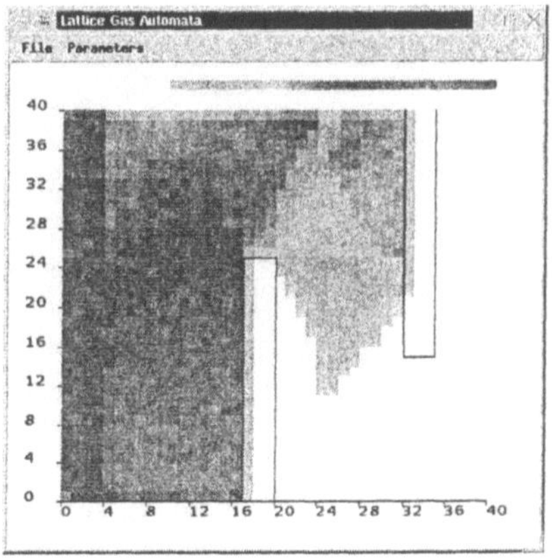

Figure 9.15. animation continued (mass density)

```
        System.out.println("random number = " + rand.nextFloat());

        XSize = 400;        // number of nodes in x
        YSize = 400;        // number of nodes in y
        iTime = 100000;     // number of time steps
        ifreq = 10;         // plot every ifreq time steps
        Domx  = 10;         // domain size in x
        Domy  = 10;         // domain size in y
        p_left = 0.3;       // going left probability
        p_right= 0.6;       // going right probability
}       // end of LGA class constructor

public void run() {
        int i=0;
```

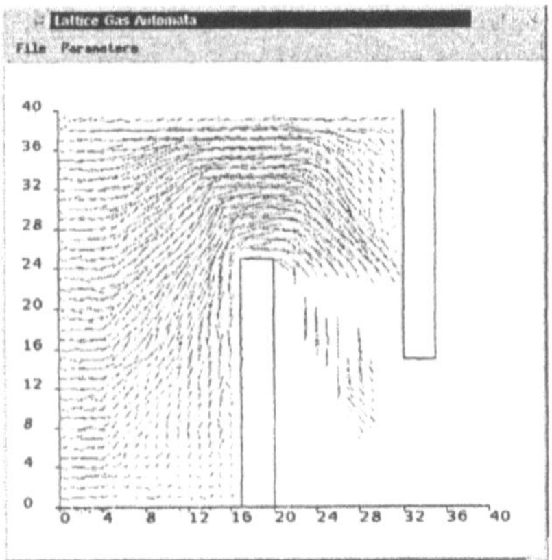

Figure 9.16. animation continued (velocity field)

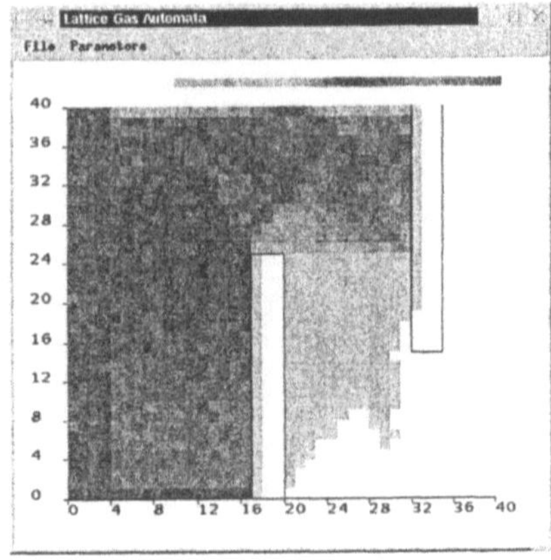

Figure 9.17. animation continued (mass density)

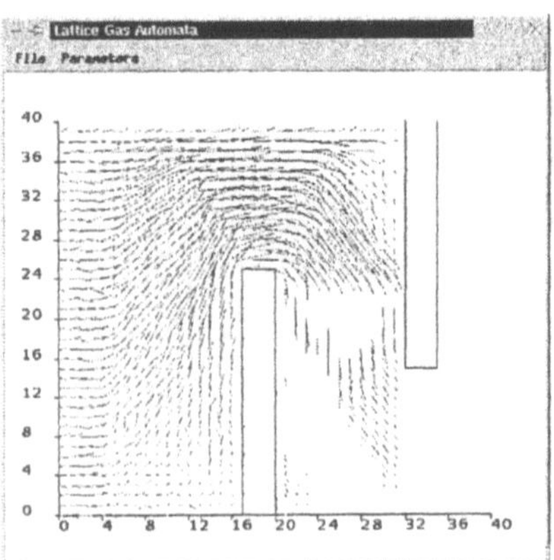

Figure 9.18. animation continued (velocity field)

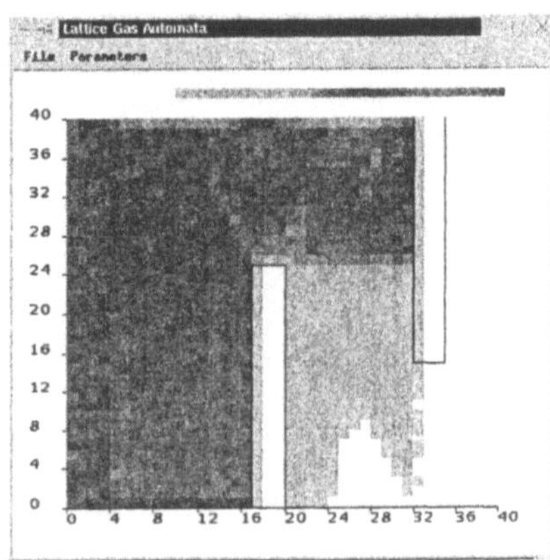

Figure 9.19. animation continued (mass density)

```
        Setup();
        parent.parent.db.DrawVelocity(Vx,Vy,mass);
        for (i=0; i<iTime; i++) {
            Inject();
            Collide();
//            Gradient(); // see the comment in the method
            Transport();
            if (i%ifreq == 0) {
               Statistics();
               parent.parent.db.DrawVelocity(Vx,Vy,mass);
            }
        }
 }
```

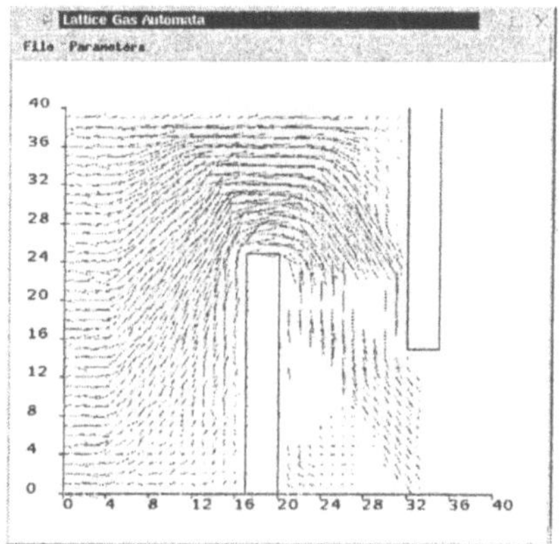

Figure 9.20. animation continued (velocity field)

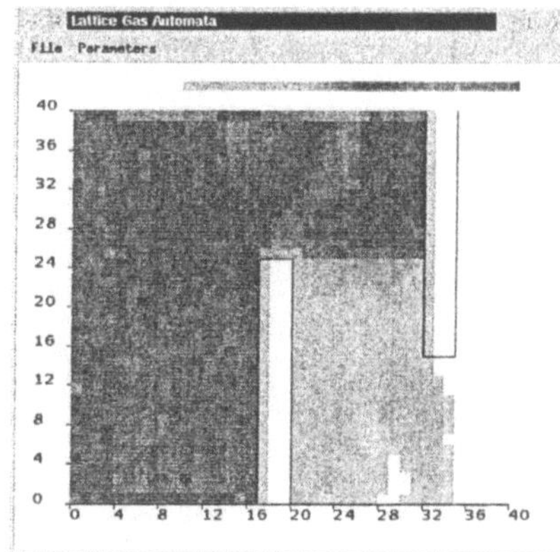

Figure 9.21. animation continued (mass density)

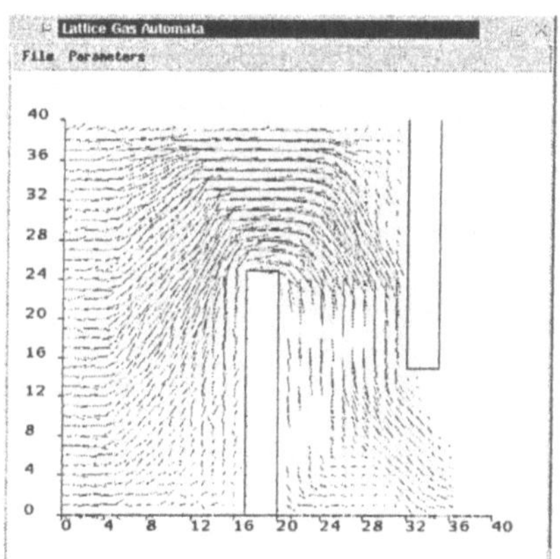

Figure 9.22. animation continued (velocity field)

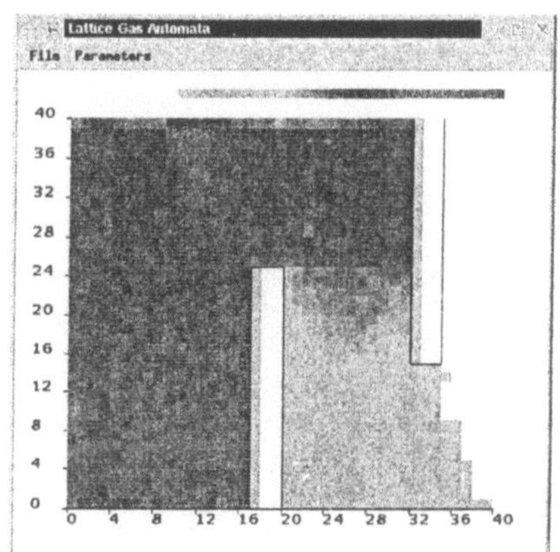

Figure 9.23. animation continued (mass density)

```
private void Setup() {
        int i,j,k;

        rand.setSeed(rand.nextLong());

        Init_m = 0;

        // tunnel (wall) boundaries
        for (i=0; i<XSize; i++) {
             state[i][1] = (byte) (1<<6);
             state[i][YSize-2] = (byte) 64;
        }

        // the following defines the obstructions
        for (i=170; i<=200; i++) state[i][250] = (byte) 64;
```

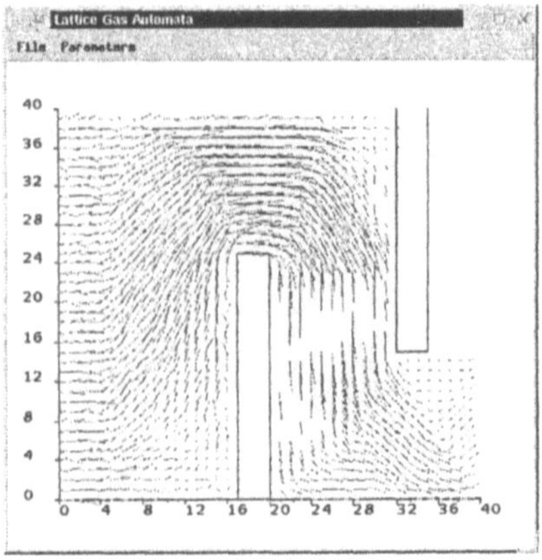

Figure 9.24. animation continued (velocity field)

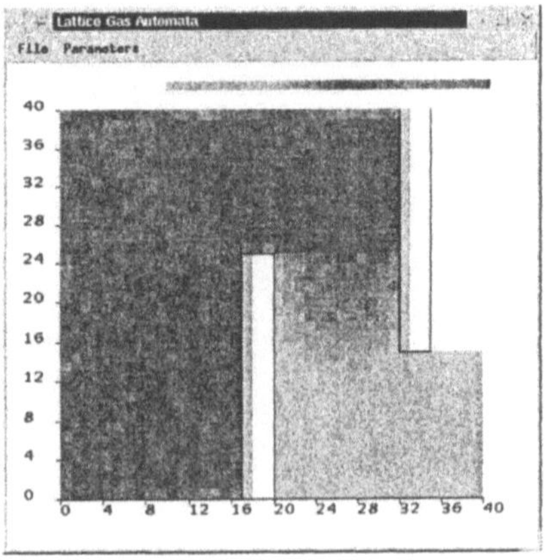

Figure 9.25. animation continued (mass density)

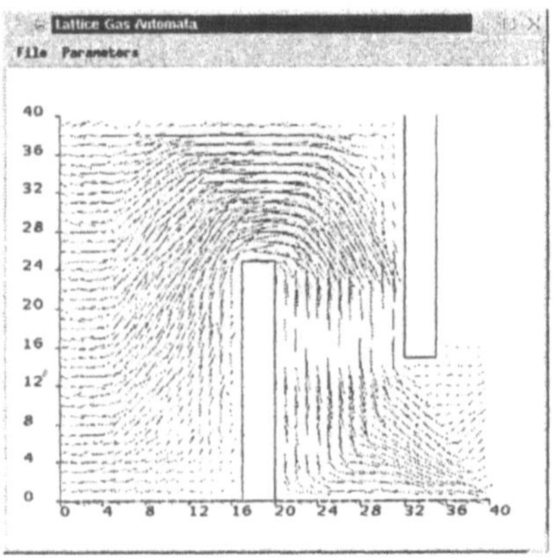

Figure 9.26. animation continued (velocity field)

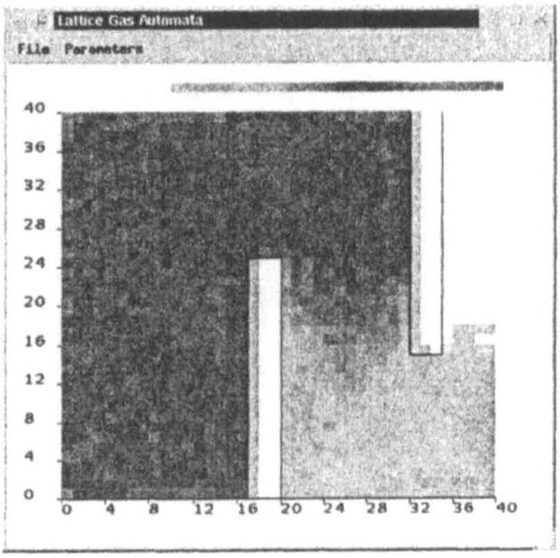

Figure 9.27. animation continued (mass density)

```
for (j=0; j<=250; j++) {
    state[170][j] = (byte) 64;
    state[200][j] = (byte) 64;
}
for (i=(170+1); i<200; i++)
    for (j=1; j<250; j++)
        state[i][j] = (byte) 0;

for (i=320; i<=350; i++) state[i][150] = (byte) 64;
for (j=150; j<400; j++) {
    state[320][j] = (byte) 64;
    state[350][j] = (byte) 64;
}
for (i=321; i<350; i++)
```

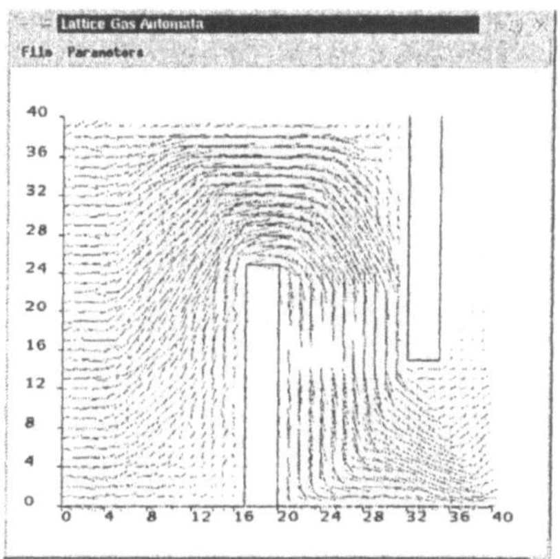

Figure 9.28. animation continued (velocity field)

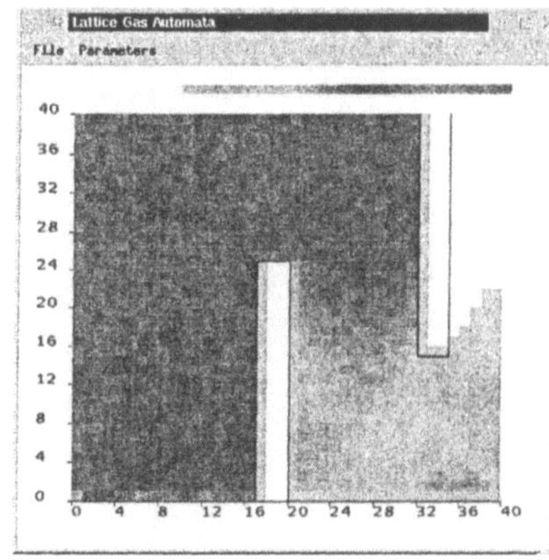

Figure 9.29. animation continued (mass density)

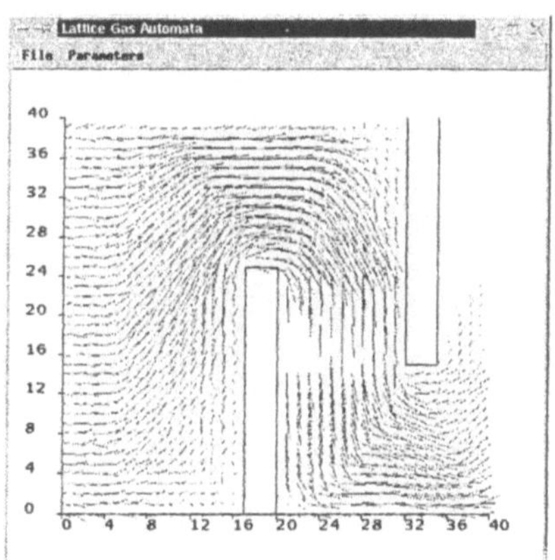

Figure 9.30. animation continued (velocity field)

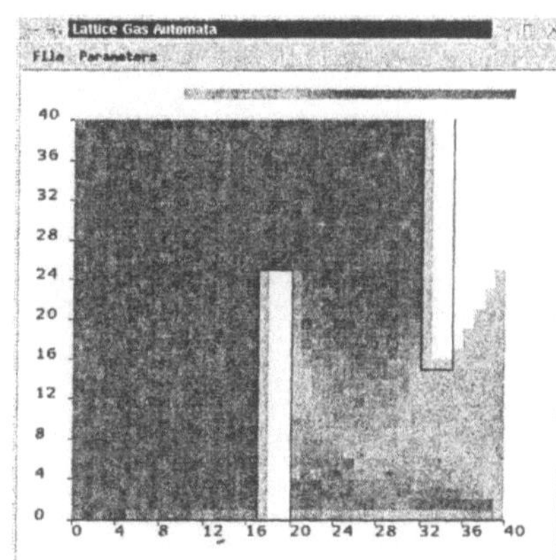

Figure 9.31. animation continued (mass density)

```
            for (j=151; j<399; j++)
                state[i][j] = (byte) 0;
}

private void Collide() {        // implements the collision rules
        byte site, bit;

        for (int i=0; i<XSize; i++) {
          for (int j=0; j<YSize; j++) {
              site = state[i][j];

              // the 'no-slip' boundary conditions: Figure 9.2 (e)
              if (site > 64) {
                 state[i][j] = (byte) 64;
```

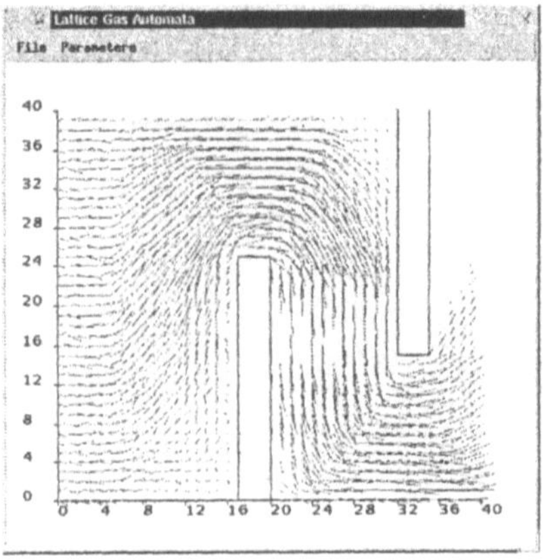

Figure 9.32. animation continued (velocity field)

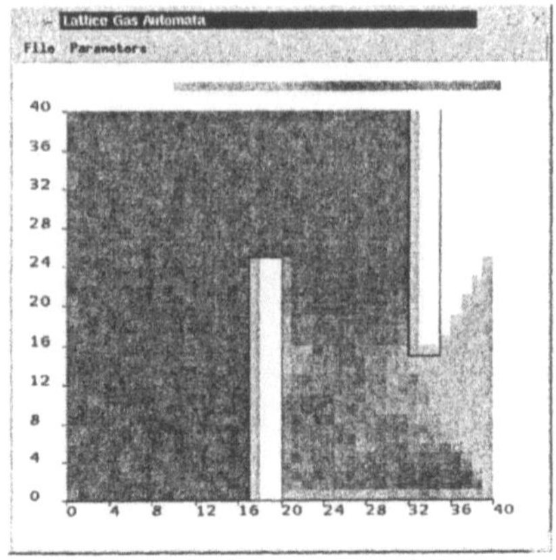

Figure 9.33. animation continued (mass density)

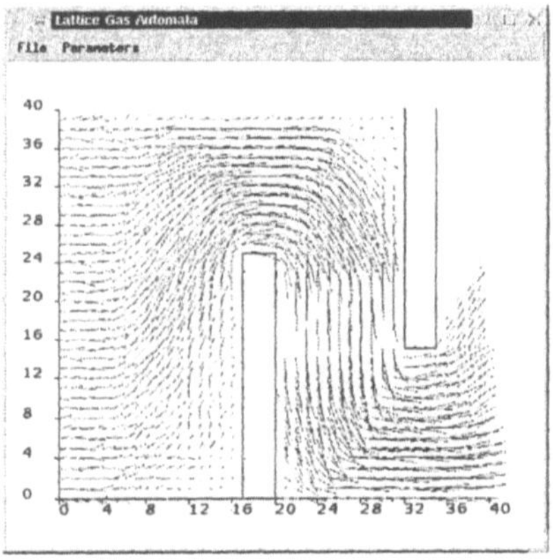

Figure 9.34. animation continued (velocity field)

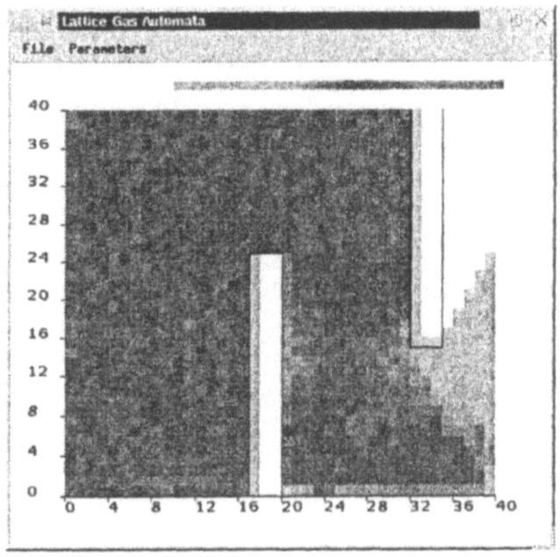

Figure 9.35. animation continued (mass density)

```
        for (int k=0; k<6; k++) {
            bit = (byte) (1 << k);
            if (((site & bit)>>>k) == 1)
                  state[i][j] |= (1 <<((k+3)%6));
        }
    } else {
        switch (site) {
         // rules of Figure 9.2 (a)
         case 9:
          if (rand.nextDouble()<0.5) state[i][j]=(byte)18;
          else state[i][j] = (byte) 36;
          break;
         case 18:
          if (rand.nextDouble()<0.5) state[i][j]=(byte)9;
```

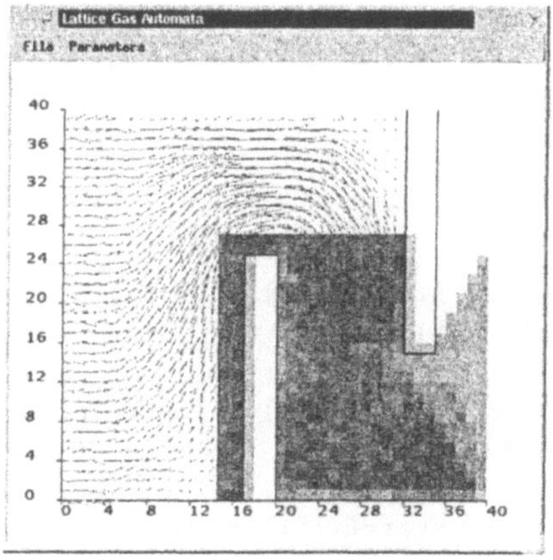

Figure 9.36. animation continued (velocity and mass)

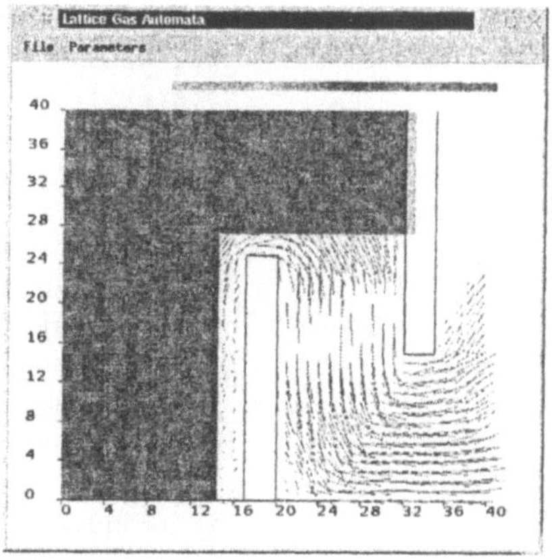

Figure 9.37. animation finished (velocity and mass)

```
   else state[i][j] = (byte) 36;
   break;
  case 36:
   if (rand.nextDouble()<0.5) state[i][j]=(byte)9;
   else state[i][j] = (byte) 18;
   break;
  // rules of Figure 9.2 (b)
  case 21:
   state[i][j] = (byte) 42;
   break;
  case 42:
   state[i][j] = (byte) 21;
   break;
  // rules of Figure 9.2 (c)
  case 50:
   state[i][j] = (byte) 41;
   break;
  case 19:
   state[i][j] = (byte) 37;
   break;
  case 22:
   state[i][j] = (byte) 13;
   break;
  case 26:
   state[i][j] = (byte) 44;
   break;
  case 37:
   state[i][j] = (byte) 19;
   break;
  case 38:
   state[i][j] = (byte) 11;
   break;
  case 44:
   state[i][j] = (byte) 26;
   break;
  case 52:
   state[i][j] = (byte) 25;
   break;
  case 11:
   state[i][j] = (byte) 38;
   break;
  case 13:
```

```
                  state[i][j] = (byte) 22;
                  break;
                case 25:
                  state[i][j] = (byte) 52;
                  break;
                case 41:
                  state[i][j] = (byte) 50;
                  break;
                // rules of Figure 9.2 (d)
                case 27:
                  if (rand.nextDouble()<0.5) state[i][j] = (byte)45;
                  else state[i][j] = (byte) 54;
                  break;
                case 54:
                  if (rand.nextDouble()<0.5) state[i][j] = (byte)27;
                  else state[i][j] = (byte) 45;
                  break;
                case 45:
                  if (rand.nextDouble()<0.5) state[i][j] = (byte)54;
                  else state[i][j] = (byte) 27;
                  break;

                default:
              } // switch
            }     // if else
        }   // inner for
      }     // outer for
}           // end of Collide

private void Transport() {
        byte site, bit;

        for (int i=0; i<XSize; i++)
          for (int j=0; j<YSize; j++)
              state_tmp[i][j] = (byte) 0;

        for (int j=1; j<YSize-1; j++) {
            site = (byte) 0;
            for (int k=0; k<3; k++) {
                bit = (byte) (1<<k);
                if (((state[XSize-2][j] & bit)>>>k) == 1) site |= bit;
            }
            if (site != 0) {
               state_tmp[2][j]  |= site;
               state_tmp[XSize-2][j] = (byte)(state[XSize-2][j]^site);
               state[XSize-2][j] ^= site;
            }
        }

        for (int i=1; i<XSize-1; i++) {
          for (int j=1; j<YSize-2; j+=2) {
              // even row
              state_tmp[i+1][j+1]  |= (state[i][j]   & (byte) 0x01);
              state_tmp[i+1][j]    |= (state[i][j]   & (byte) 0x02);
              state_tmp[i+1][j-1]  |= (state[i][j]   & (byte) 0x04);
              state_tmp[i][j-1]    |= (state[i][j]   & (byte) 0x08);
              state_tmp[i-1][j]    |= (state[i][j]   & (byte) 0x10);
              state_tmp[i][j+1]    |= (state[i][j]   & (byte) 0x20);
              // odd row
              state_tmp[i][j+2]    |= (state[i][j+1] & (byte) 0x01);
              state_tmp[i+1][j+1]  |= (state[i][j+1] & (byte) 0x02);
              state_tmp[i][j]      |= (state[i][j+1] & (byte) 0x04);
              state_tmp[i-1][j]    |= (state[i][j+1] & (byte) 0x08);
              state_tmp[i-1][j+1]  |= (state[i][j+1] & (byte) 0x10);
              state_tmp[i-1][j+2]  |= (state[i][j+1] & (byte) 0x20);
          }
```

```
        }

        for (int i=0; i<XSize; i++) {
          for (int j=0; j<YSize; j++) {
               if (state[i][j] < 0) System.out.println("negative");
               if (state[i][j] >= 64) {
                  state[i][j] =(byte)(64+state_tmp[i][j]);
               } else state[i][j] = state_tmp[i][j];
          }
        }
 }      // end of Transport

 private void Statistics() {  // coarse grained quantities
        byte bit;
        int va, vb, vc, vd, ve, vf, itmp;
        int ii, jj;

        current_m = 0;
        for (int i=0; i<XSize/Domx; i++) {
          for (int j=0; j<YSize/Domy; j++) {
               va = 0;
               vb = 0;
               vc = 0;
               vd = 0;
               ve = 0;
               vf = 0;
               itmp = 0;
               for (int m=0; m<Domx; m++) {
                 for (int n=0; n<Domy; n++) {
                      ii = m+i*Domx;
                      jj = n+j*Domy;
                      va += ((state[ii][jj] & (byte) 1)>>>0);
                      vb += ((state[ii][jj] & (byte) 2)>>>1);
                      vc += ((state[ii][jj] & (byte) 4)>>>2);
                      vd += ((state[ii][jj] & (byte) 8)>>>3);
                      ve += ((state[ii][jj] & (byte) 16)>>>4);
                      vf += ((state[ii][jj] & (byte) 32)>>>5);
                 }    // n loop
               }      // m loop
               Vx[i][j] = (va+vc-vd-vf)*Math.cos(Math.PI/3.0)+(vb-ve);
               Vy[i][j] = (va+vf-vc-vd)*Math.cos(Math.PI/6.0);
               Vx[i][j] /= (Domx*Domy);
               Vy[i][j] /= (Domx*Domy);
               Vx[i][j] *= 2.0;  // the factor of 2 is solely
               Vy[i][j] *= 2.0;  // for the purpose of plotting
               itmp = va+vb+vc+vd+ve+vf;
               mass[i][j] = ((double) itmp)/Domx/Domy;
               current_m += itmp;
          }    // j loop
        }      // i loop
        System.out.println("density = "+
                              ((double)current_m/XSize/YSize));
 }      // end of method Statistics

/*
 private void Gradient() {
 // you can put effects such as gravity here
 }
*/
 private void Inject() {
        int i,j,k;
        double r;

        r = ((double) current_m)/XSize/YSize;
```

```
            if (r < 3.0) {           // unphysical otherwise
               for (j=1; j<YSize-1; j++) {
                  for (i=1; i<(1+(Domx+Domy)*2); i++) {
                     state[i][j] = (byte) 0;
                     for (k=0; k<3; k++)
                        if (rand.nextDouble() < p_right)
                           state[i][j] |= (byte)(1<<k);
                     for (k=3; k<6; k++)
                        if (rand.nextDouble() < p_left)
                           state[i][j] |= (byte)(1<<k);
                  }
               }
            }
   }        // end of Inject
}           // end of LGA class
```

Listing 9.1 LGA.java

Figures 9.4 to 9.37 animate the gas flow when the program is running. Black lines outline the obstructions, which can be imagined as walls in a hallway (top view). The space is initially empty and the gas enters into the space from the left at a constant rate. The left column of the plots show the (domain averaged) velocity fields while the left column are the (domain averaged) mass densities. The color bar on the top of the mass plot linearly maps the mass density of the domain. The densest color is red, representing an average of 4.5 molecules per lattice node in that domain. Plotting is performed at a rate of every 10 time steps. The animation switches between velocity plot and mass plot. In the end of the animation (Figure 9.37), the gas flow reaches a steady state.

9.6 Summary

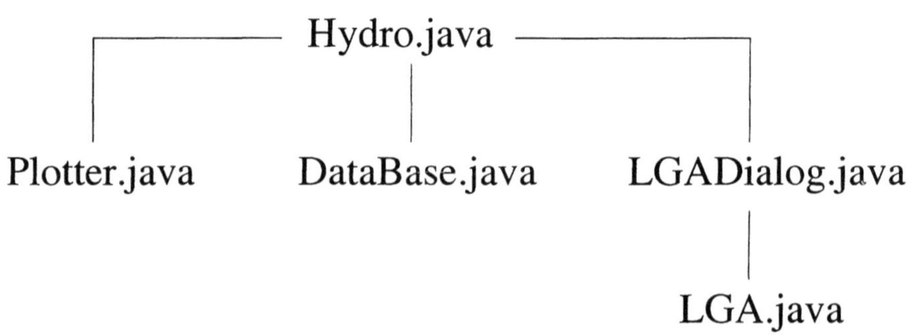

Figure 9.38. Source programs for the lattice gas automata simulation

The `main()` method, plotting, and dialog classes can be easily written using as templates classes in previous chapters.

We presented a simulation of (low Reynolds number) 2-dimensional air flows by lattice gas automata. Since the method of cellular automata implements sound microscopic laws in the updating rules, the simulation reveals much of the macroscopic behavior of the system when suitably defined quantities are calculated. The method is easily modified for complex and irregular boundaries.

Nature exhibits an infinite number of examples of seemingly complex patterns, phenomena, and behaviors. However, the operating rules behind the complexity can sometimes be disproportionately simple. In a minimalist's viewpoint, it is the simple underlying rules that govern the way nature works. Cellular automata simulation on computers is a reverse-engineering means to unravel the simple rules.

9.7 References and Further Reading

A treatise on complexity is, R. Badii and A. Politi, "Complexity - Hierarchical structures and scaling in physics", Cambridge University Press (1997)

Various applications of cellular automata can be found in, S. Wolfram, "Cellular Automata and Complexity", Addison-Wesley, Reading, MA (1994), B. Chopard and M. Droz, "Cellular Automata Modeling of Physical Systems", Cambridge University Press (1998), and S. Wolfram, "A New Kind of Science", Wolfram Media, Inc. IL (2002)

Self-organized criticality was described in, P. Bak, C. Tang, and K. Wiesenfeld, "Self-Organized Criticality: An Explanation of 1/f Noise", *Phys. Rev. Lett.*, **59** (1987) 381-384

2- and 3-dimensional lattice gas automata were demonstrated in, U. Frisch, B. Hasslacher, and Y. Pomeau, "Lattice-Gas Automata for the Navier-Stokes Equation", *Phys. Rev. Lett.* **56** (1986) 1505 S. Wolfram, "Cellular Automaton Fluids I. Basic Theory", *J. Stat. Phys.*, **45** (1986) 471, and U. Frisch et al., "Lattice-Gas Hydrodynamics in Two and Three Dimensions", *Complex Syst.*, **1** (1987) 648

Santa Fe Institute actively engages in researches on complex systems and emerging science. Its web site is worth visiting on a regular basis, `www.santafe.edu`

Chapter 10

PATH INTEGRAL

Feynman's method of path integration offers an alternative to the conventional solutions to the Schrodinger's equation. Path integrals provide not only a new computational approach to quantum mechanics, but also a different conceptual perspective of view. The advantage of path integral manifests itself particularly when the number of particles (or number of degrees of freedom) of the many-body system increases. Furthermore, the formalism derived for the dynamics of a system can, after slight modification, be applied to calculate interesting quantities of systems in thermodynamic equilibrium.

A plethora of applications of path integral can be found in chemical physics, statistical mechanics, condensed matter physics, nuclear and particle physics. To lengthen the list, we in this chapter demonstrate its application to financial engineering.

10.1 Feynman's Sum Over Histories

The energy, H, of a (non-relativistic) many-body system consists of kinetic, T, and potential energy, V,

$$H = T + V = \sum_{i=1}^{n} \frac{p_i}{2m} + V(x_1, x_2, \cdots, x_n), \tag{10.1}$$

where p and m are the momentum and mass of the (spinless) particle, respectively, n is the total number of particles in the system. Note that because of the Heisenberg uncertainty principle of quantum mechanics (position and velocity can not be precisely measured *at the same time*), p and x do not commute. In other words,

$$[p_i, x_j] \equiv p_i x_j - x_j p_i = -i\hbar\delta_{ij}, \tag{10.2}$$

where $\delta_{ij} = 1$ if $i = j$, and 0 otherwise. $\hbar$ is Planck's constant, h, divided by 2π with $h = 6.626 \times 10^{-34}$ joule second. Planck's constant characterizes the scale (or importance) of quantum effects. If the value of $\hbar$ can be ignored and therefore approximated as zero, we return to the classical world. [1] The state of the system is described by the wavefunction, $\Psi(x_1, x_2, \cdots, x_n, t) = \Psi(x, t) = \langle x|\Psi(t)\rangle$, whose absolute value squared gives the probability that, at time t, particle i is located at position x_i, and so on. $|x\rangle$ ($|p\rangle$) means *in coordinate (momentum) representation*, and $\langle x|$ ($\langle p|$) is the complex conjugate of $|x\rangle$ ($|p\rangle$). Dynamics of the system is governed by the Hamiltonian H of Eq. (10.1),

$$H\Psi = i\hbar\frac{\partial\Psi}{\partial t}. \tag{10.3}$$

From Eq. (10.3), it is seen that the wavefunction, initially spreading over the region, x_0, equals, at time t,

$$\begin{aligned}\Psi(x,t) = \langle x|\Psi(t)\rangle &= \langle x|e^{-iHt/\hbar}\Psi(0)\rangle \\ &= \int dx_0\, \langle x|e^{-iHt/\hbar}|x_0\rangle\langle x_0|\Psi(0)\rangle \\ &= \int dx_0\, \langle x|e^{-iHt/\hbar}|x_0\rangle\Psi(x_0,0) \\ &= \int dx_0\, K(x,x_0;t)\Psi(x_0,0),\end{aligned} \tag{10.4}$$

where we have used the completeness property of position eigenstates,[2]

$$\int dx\, |x\rangle\langle x| = 1. \tag{10.5}$$

K in Eq. (10.4) is called *propagator* of the system. If the propagator is known, the dynamics of the system is solved. Let's therefore take a closer look at it.

We now divide the time interval between 0 and t into N slices. Each slice of the time interval, Δt, equal to t/N. Next, we observe that the time evolution operator, K, can be factorized into a product of N operators, each evolving a short time interval Δt, according to the following formula,

$$e^{-iHt/\hbar} = e^{-iH\Delta t/\hbar} \cdot e^{-iH\Delta t/\hbar} \cdots e^{-iH\Delta t/\hbar}. \tag{10.6}$$

Together with a repeated use of the identity operator of Eq. (10.5), we can express the (finite-time) propagator as a product of short-time propagators,

[1] Recent experiments have shown that trillions of atoms as a whole can be prepared in a quantum mechanically entangled state. Quantum effects therefore depend not on the scale of the system but on how well the system is isolated from outside disturbances.

[2] Eigenstates are component functions of which a wavefunction can be represented as a (weighted) sum. For this to be true, eigenfunctions have to carry a couple of properties. Eq. (10.5) is one of them.

$K(x_k, x_{k-1}; \Delta t)$,

$$K(x, x_0; t) = \int dx_1 \int dx_2 \cdots \int dx_{N-1} \prod_{k=1}^{N} K(x_k, x_{k-1}; \Delta t), \tag{10.7}$$

where

$$K(x_k, x_{k-1}; \Delta t) = \langle x_k | e^{-iH\Delta t/\hbar} | x_{k-1} \rangle. \tag{10.8}$$

The formulas have so far been exact. To proceed, we need an approximation, namely,

$$\begin{aligned} e^{-iH\Delta/\hbar} &= e^{-i(T+V)\Delta t/\hbar} = e^{-iT\Delta t/\hbar} e^{-iV\Delta t/\hbar} + O(\Delta t^2) \\ &\approx e^{-iT\Delta t/\hbar} e^{-iV\Delta t/\hbar}. \end{aligned} \tag{10.9}$$

$e^{-iV\Delta/\hbar}$ can simply be moved out of the bracket in Eq. (10.8) since it commutes with position eigenstates. H is left with T in the short-time propagator. Now the orthonormal conditions of the momentum eigenstates read,

$$\int dp\, |p\rangle\langle p| = 1 \text{ and } \langle p_i | x_j \rangle = (2\pi\hbar)^{-1/2} e^{-ip_i x_j/\hbar} \delta_{ij}. \tag{10.10}$$

We have,

$$\begin{aligned} K(x, x_0; \Delta t) &\approx e^{-iV\Delta t/\hbar} \int dp\, \langle x | e^{-iT\Delta t/\hbar} | p \rangle \langle p | x_0 \rangle \\ &= \frac{e^{-iV\Delta t/\hbar}}{(2\pi\hbar)^n} \int_{-\infty}^{\infty} dp\, e^{-i\Delta t p^2/2m\hbar} e^{ip(x-x_0)/\hbar} \\ &= \left[\frac{m}{2\pi i\hbar\Delta t}\right]^{n/2} e^{\frac{i}{\hbar}\left(\frac{m}{2\Delta t}|x-x_0|^2 - V(x)\Delta t\right)}. \end{aligned} \tag{10.11}$$

The finite-time propagator then becomes,

$$\begin{aligned} K(x, x_0; t) \approx{}& \left[\frac{mN}{2\pi i\hbar t}\right]^{nN/2} \int dx_1 \cdots \int dx_{N-1} \\ & \times \exp\left\{\frac{i}{\hbar}\left[\frac{mN}{2t}\sum_{k=1}^{N} |x_k - x_{k-1}|^2 - \frac{t}{N}\sum_{k=1}^{N} V(x_k)\right]\right\}. \end{aligned} \tag{10.12}$$

The quantity in the square parentheses of the exponent is the classical *action* along the paths between x_0 and x. The propagator is thus a weighted sum of paths (or histories) between two fixed end points, x_0 and x,

$$K(x, x_0; t) \propto \sum_{x(t)} e^{\frac{i}{\hbar} S[x(t)]}, \tag{10.13}$$

where the action, S, is defined as,

$$S[x(t)] = \int_0^t dt' \left[\frac{m}{2}\dot{x}(t')^2 - V\left(x(t')\right)\right] = \int_0^t dt' \mathcal{L}, \tag{10.14}$$

and $\mathcal{L}$ is the classical Lagrangian of the system.

10.2 Numerical Path Integration and Feynman-Kac Formula

It is reminded that the probability is the square of the absolute value of the wavefunction Ψ (or the propagator). Cancellation and interference can arise for paths whose actions differ by an amount greater than $\hbar$, due to the highly oscillatory behavior of the phase. Therefore, it is clear that most terms in the sum do not contribute except those whose values of action remain within $\hbar$ over the region of the paths,

$$\frac{\delta S[x(t)]}{\delta x(t)} \leq \hbar. \tag{10.15}$$

The so-called sign problem has hindered efficient numerical evaluations of the (real time) propagator. We note, again, that if $\hbar$ is too small to be significant in the problem under consideration, classical equations of motion are re-derived from the condition of stationary action,

$$\frac{\delta S[x(t)]}{\delta x(t)} = 0. \tag{10.16}$$

If we let $x_0 = x$ in the propagator and integrate over all x's, we get,

$$\int dx\, K(x, x; t) = \sum_k e^{-iE_k t/\hbar}, \tag{10.17}$$

where E_k are the eigenvalues (energy levels) of the system. If we substitute an imaginary time $t = -i\hbar\tau$ in Eq. (10.17), the sum is, for large positive τ, dominated by the contribution from the ground state (state of the lowest energy), $e^{-\tau E_0}$. We therefore obtain the important result,

$$E_0 = -\lim_{\tau\to\infty} \frac{1}{\tau} \ln \int dx K(x, x; -i\hbar\tau), \tag{10.18}$$

which allows us to evaluate the ground state energy of the system without knowing its wavefunction. By the use of the imaginary time, the scourge of rapid oscillations (cancellations) disappears and numerical integrations become feasible. Equation (10.18) is called Feynman-Kac formula and lends itself to applications where wavefunction solutions are intractable.

To demonstrate how to sum Feynman's paths on a computer, we take the pricing of financial options as an example in following sections.

10.3 Options in Finance

Suppose you are considering a coupon which entitles you to buy 50 liters of gasoline for K dollars (called strike price) a month from today. You know of today's gas price. If the price a month later soars above K, the coupon really pays you off. If, on the other hand, the gas price plunges below K, the coupon ends up worthless. The question now is, What is the fair price of the coupon for you to buy? The coupon can be a nice deal for those whose pockets are thin and are, for instance, planning a long trip by auto next month. If the gas price next month does rise, the savings can be significant. If, however, the price drops, the cost is merely what is paid for the coupon. The risk of volatile gas prices the buyers (drivers) are exposed to is shifted to coupon sellers. To neutralize the risk, the sellers can hedge by setting up a portfolio which consists of, for example, selling coupons and buying a certain quantity of gas.

Similar activities routinely take place in financial markets, where an *option* (for example, our coupon) is a financial instrument whose value is contingent on an underlying asset (gas), such as the stock price of a company. The price f of an option is a function of the time to maturity T, the current price of the underlying asset S_0, and the strike price K. It is described by the famous Black-Scholes-Merton equation,

$$\frac{\partial f}{\partial t} = -rS\frac{\partial f}{\partial S} - \frac{1}{2}\sigma^2 S^2 \frac{\partial^2 f}{\partial S^2} + rf, \qquad (10.19)$$

where r is the risk-free interest rate and σ is the volatility of the underlying asset S. Our task is to solve Eq. (10.19) by path integral method.

10.4 A Path Integral Approach to Option Pricing

Since asset prices S can not be negative, a change of variable, $S = e^x$ for $-\infty \leq x \leq \infty$ takes the Black-Scholes-Merton equation into,

$$\frac{\partial f}{\partial t} = \Big[(\frac{\sigma^2}{2} - r)\frac{\partial}{\partial x} - \frac{\sigma^2}{2}\frac{\partial^2}{\partial x^2} + r\Big] f = (H_{BSM} + r)f, \qquad (10.20)$$

where H_{BSM} is defined as the 'Hamiltonian' of the portfolio.

Now since the distribution of the values of the option at maturity is known to be,

$$\begin{aligned} f(T, S(T)) &= \begin{cases} S(T) - K, & S(T) \geq K \\ 0, & S(T) < K \end{cases} \\ &\equiv g(S(T)), \end{aligned} \qquad (10.21)$$

the value of the option at any time t ($t < T$) is related to $g(S(T))$ via the propagator, $K_{BSM}(x, x'; t)$,

$$\begin{aligned} f(t,x) &= e^{-r(t-T)} \int_{-\infty}^{\infty} dx' \; K_{BSM}(x,x';t) g(x') \\ &= e^{-r(t-T)} \int_{-\infty}^{\infty} dx' \; \langle x | e^{-(T-t)H_{BSM}} | x' \rangle g(x'). \end{aligned} \tag{10.22}$$

Transforming the integral to the momentum space and performing the Gaussian integral, as was done in Eq. (10.11), we get,

$$\begin{aligned} K_{BSM}(x,x';t) &= \langle x | e^{-\tau H_{BSM}} | x' \rangle \\ &= \frac{1}{\sqrt{2\pi\tau}\sigma} \exp\left\{ -\frac{1}{2} \left[\frac{x - x' + \tau(r - \frac{\sigma^2}{2})}{\sqrt{\tau}\sigma} \right]^2 \right\}, \end{aligned} \tag{10.23}$$

where the time $\tau = T - t$ runs backwards, and the propagator 'evolves' the stock price at T back to the price at present. Equation (10.23) also says that the stock price logarithm at T, $\ln(S(T)) = x'$, is Gaussian distributed with a mean of $\ln(S(t)) + (r - \sigma^2/2)(T - t)$ and a variance of $\sigma^2(T - t)$. These facts were already implicit in the Black-Scholes-Merton equation. The propagator formalism simply makes it explicit. A derivation starting from the first principles [i.e., the Brownian motion (or Wiener process) of the stock price], instead of the Black-Scholes-Merton equation, is possible, but it's beyond the scope of this book.

There is not any imaginary number $i = \sqrt{-1}$ in the Black-Scholes-Merton-Schrodinger equation [Eq. (10.20)]. Furthermore, the option price function $f(T, S(t))$ is a real-valued function in contrast to the complex wavefunction (probability amplitude) in quantum mechanics. Both make the interpretation and evaluation of the path integral straightforward. Equation (10.22) states that the payoff of the option is a sum of payoffs at all possible stock prices at maturity, with individual terms in the sum weighted by the probability of occurrence of the stock price trajectory from the current stock price to the stock price at maturity. Empirical stock price distributions, other than the log-normal distribution inherent in the Black-Scholes-Merton equation, can be easily implemented under Feynman's path integral framework.

We are going to employ the following technique to perform summation of paths.

10.5 Importance Sampling (Metropolis-Hastings algorithm)

The integral of Eq. (10.22) has the general form,

$$I = \int dx \; \rho(x) g(x), \tag{10.24}$$

where $\rho(x)$ is a normalized, positive definite function of x, and $g(x)$ is a smooth function of x. If vector x has n dimensions and each dimension is divided into N slices between the end points, a brute-force evaluation of the multi-dimensional integration by summing the function values at the n by N discrete lattice points is not efficient, if not impossible.

A more efficient algorithm is to sample $g(x)$ from the distribution $\rho(x)$. Suppose x_i is generated. The next pick, x_{i+1}, is calculated from a random deviate from x_i. The deviation can have a maximum cap depending on the specific problem. We then evaluate the ratio,

$$\omega = \frac{\rho(x_{i+1})}{\rho(x_i)}. \tag{10.25}$$

If $\omega > 1$, indicating that x_{i+1} enters into a more favorable region, the move is accepted. If, however, $\omega < 1$, draw a random number ς between 0 and 1, and accept x_{i+1} if $\varsigma < \omega$. Otherwise, it's rejected. The algorithm (Metropolis-Hastings) ensures that moves are not trapped in local regions of $\rho(x)$ and that all important regions of the configuration space are sampled. An estimate of the integral of Eq. (10.24) after M evaluations is,

$$\bar{I} = \frac{1}{M}\sum_{i=1}^{M} g(x_i) = I + O(1/\sqrt{M}), \tag{10.26}$$

which says that, with the technique of importance sampling, each move is equally important. The overline means average. The error in $\bar{I}$ can be calculated in the usual way,

$$\Delta\bar{I} = \left[\frac{\overline{I^2} - (\bar{I})^2}{M}\right]^{\frac{1}{2}}. \tag{10.27}$$

An advantage of the path integral method is that other quantities of interest can be calculated in the same Monte Carlo simulation of the path. For example, the sensitivity of the option price to variable X is,

$$\begin{aligned}\frac{\partial f(x,X,t)}{\partial X} &= \\ \int dx' \left[g(x',X)\frac{\partial \ln K(x,x',X;t)}{\partial X} + \frac{\partial g(x',X)}{\partial X}\right] &K(x,x',X;t).\end{aligned} \tag{10.28}$$

Note the propagator is factored out in the above equation. The quantity in the square parentheses simply defines a new $g(x)$. By the use of the same importance sampling in the path integral Monte Carlo, the estimate (average value) of the sensitivity becomes,

$$\overline{\frac{\partial f(x,X,t)}{\partial X}} = \frac{1}{M}\sum_{i=1}^{M}\left[g(x_i,X)\frac{\partial \ln K(x,x_i,X;t)}{\partial X} + \frac{\partial g(x_i,X)}{\partial X}\right], \tag{10.29}$$

in which the derivative of the propagator logarithm is readily available as a sum of the short-time propagator logarithms,

$$\frac{\partial \ln K(x, x_i, X; t)}{\partial X} = \sum_{k=1}^{N} \frac{\partial \ln K(x_k, x_i, X; t)}{\partial X}. \tag{10.30}$$

Many interesting Greeks (sensitivities) of options can therefore be obtained after a single set of Monte Carlo simulations of the paths, as will be shown in our source program.

10.6 Implementation

To use importance sampling in this application, we need to start from a stock price path that is realistic, that it, one that looks like, for example, any one in Figures 10.2 to 10.9. Before getting there, we firstly generate a stock price movement path without the variance term ($\sigma = 0$). We then start invoking importance sampling and let the path evolve for enough time before it 'relaxes' into a realistic one. This relaxation process takes a typical of 100 Monte Carlo steps.

At each time slice, i ($i = 1, 2, \cdots, N$ and $N\Delta t$ is the maturity time), the probability distribution of Eq. (10.23) is, apart from a normalization constant,

$$\rho(i-1, i) = \exp\left[-\frac{(x_i - x_{i-1} - \mu\Delta t)^2}{2\sigma^2\Delta t}\right], \tag{10.31}$$

where $\mu = r - \sigma^2/2$. A random trial price movement x_i' is generated from x_i by,

$$x_i' = x_i + \varsigma\Delta, \tag{10.32}$$

where ς is a random number between -1 and +1, and $\Delta = \lambda\sigma\sqrt{\Delta t}$. λ is a constant to be chosen so that the rate of successful trials is about 0.5 for the sake of efficiency. We now proceed with the importance sampling for this time slice, calculating the ratio of $\rho(i-1, i')$ to $\rho(i-1, i)$. After finishing all the N slices, we have a new path. The payoff function and other sensitivity derivatives can be calculated and accumulated. This constitutes one Monte Carlo path. After, say, 10^5 Monte Carlo paths, the means and their errors can be obtained.

Listing 10.1 shows the class implementing path integration with importance sampling. Figure 10.1 shows the dialog box for user interaction with the options by path integral application.

```
/*      Sun-Chong Wang
        TRIUMF
        4004 Wesbrook Mall
        Vancouver, V6T 2A3
        Canada
```

```
        e-mail: wangsc@triumf.ca

        Path.java generates paths by importance sampling */

import java.lang.*;
import java.util.*;
import java.util.Random;

public class Path implements Runnable {
 Random rand;
 int iRelax;
 int ifreq;
 static int nSteps;
 static int nTrials;
 static double S;
 static double dt, mu, variance, X, r;
 double Payoff, Delta, Kappa, Rho;
 double Payoff_error, Delta_error, Kappa_error, Rho_error;
 double lambda, shift;
 double[] path;

 PIDialog parent;

 public Path(PIDialog parent) {
        this.parent = parent;
        rand     = new Random();
        lambda   = 2.5;
        shift    = 0.0;
        iRelax   = 100;
        ifreq    = 1000;
        Payoff_error = 0.0;
        Payoff   = 0.0;
        System.out.println("random number = " + rand.nextFloat());
 }      // end of Path class constructor
```

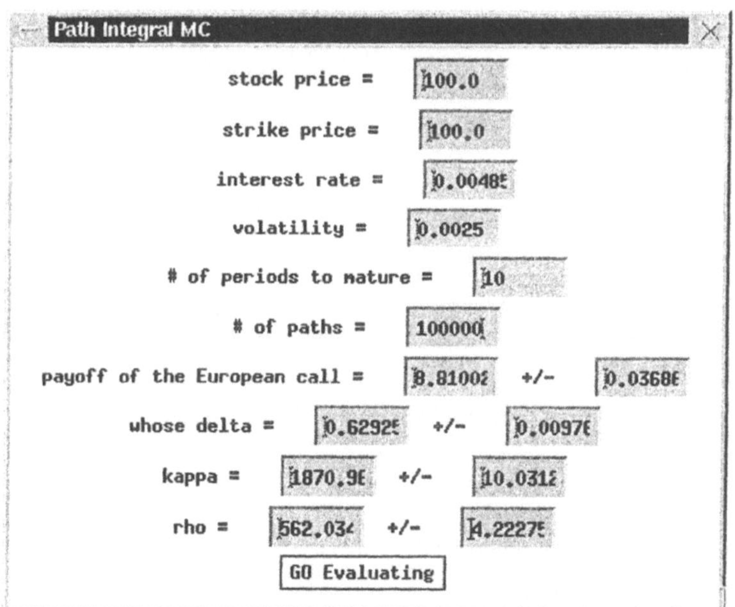

Figure 10.1. Dialog box holding parameters for the path integral Monte Carlo

Figure 10.2. Possible stock price movements by path integral Monte Carlo

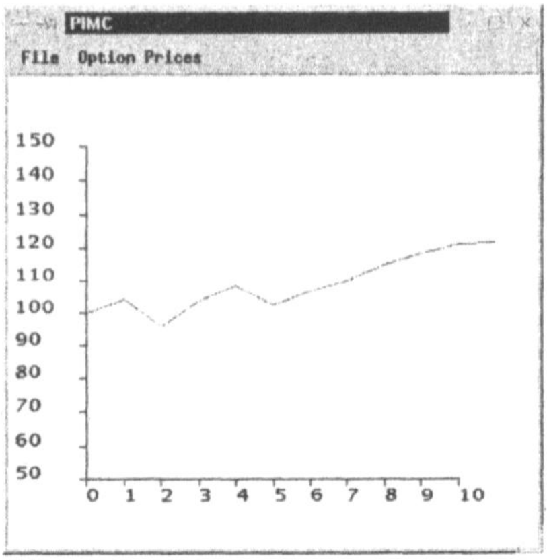

Figure 10.3. Possible stock price movements by path integral Monte Carlo

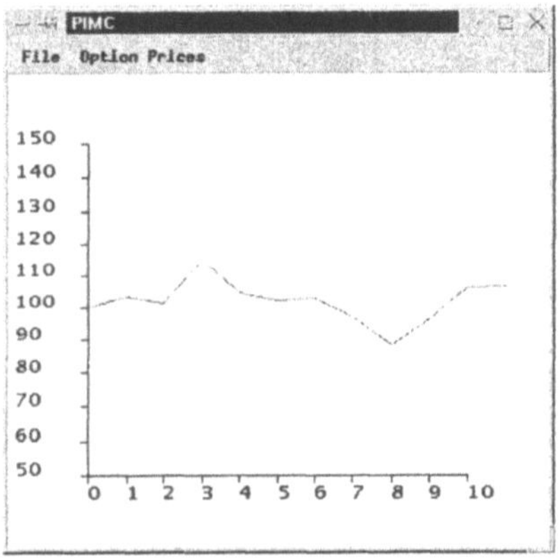

Figure 10.4. Possible stock price movements by path integral Monte Carlo

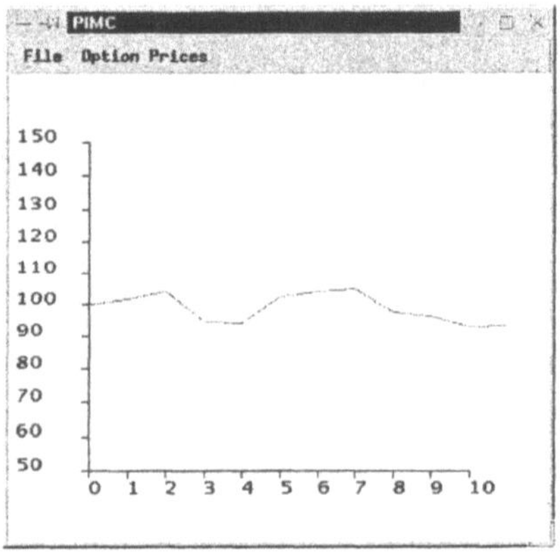

Figure 10.5. Possible stock price movements by path integral Monte Carlo

```
public void run() { // this method is run by a thread
        Initialize();
        Mean();
}

public void Initialize() {
        double[] y = new double[nSteps+2];

        mu = r - variance/2.0;
        dt = 1.0;
        y[0] = Math.log(S);  // stock price logarithm
        for (int i=1; i<=(nSteps+1); i++) {
            y[i] = y[i-1] + mu*dt;
        }   // this is the initial deterministic path
```

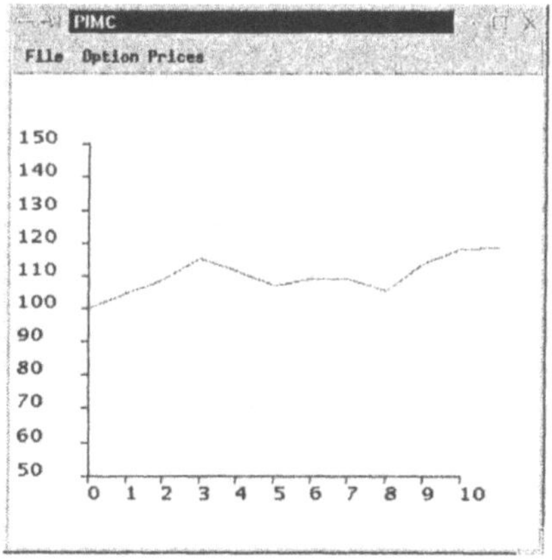

Figure 10.6. Possible stock price movements by path integral Monte Carlo

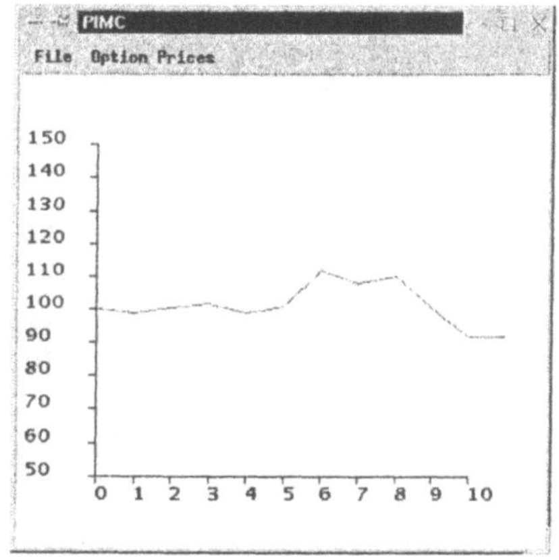

Figure 10.7. Possible stock price movements by path integral Monte Carlo

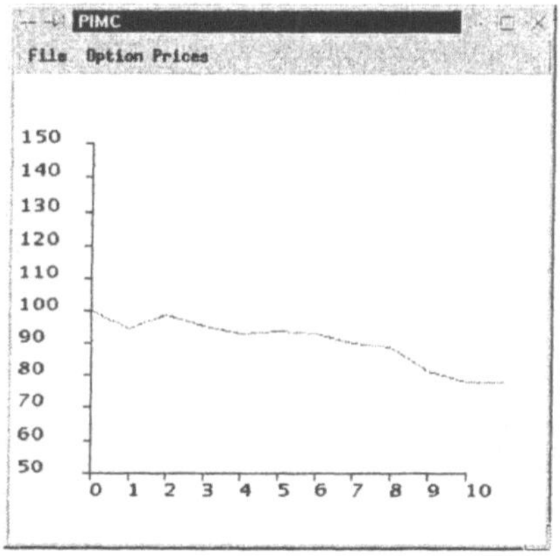

Figure 10.8. Possible stock price movements by path integral Monte Carlo

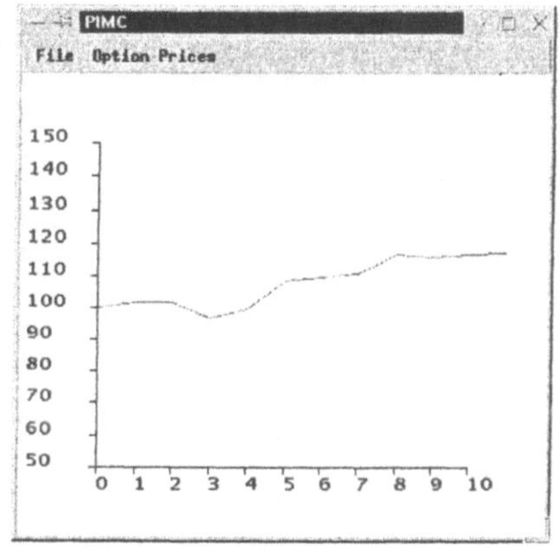

Figure 10.9. Possible stock price movements by path integral Monte Carlo

```
        path = y;
        rand.setSeed(rand.nextLong());
        shift = lambda*Math.sqrt(variance*dt);
}

public void Mean() {
        int M;
        double F, F2, F_d, F_d2, F_k, F_k2, F_i, F_i2, S_T, tmp;
        double Payoff2, Delta2, Kappa2, Rho2;

        // relax from the deterministic (unrealistic) path
        for (int i=0; i<iRelax; i++) shake();

        M = 0;
```

```
        Payoff = 0.0;
        Payoff2 = 0.0;
        Delta = 0.0;
        Delta2 = 0.0;
        Kappa = 0.0;
        Kappa2 = 0.0;
        Rho = 0.0;
        Rho2 = 0.0;
        do {         // the following performs the summation
           M += 1;  // in Eqs. (10.26), (10.27), (10.29)
           shake();
           // parent of this class is PIDialog, whose parent is Options
           // (cf. Figure 10.10) in which db is defined as an instance
           // of DataBase
           if ((M-1)%ifreq == 0) parent.parent.db.pathToDraw(path);
           S_T = Math.exp(path[nSteps]);
           F = Math.max(S_T - X, 0.0)*Math.exp(-r*nSteps*dt);
           F2 = F*F;
           Payoff += F;
           Payoff2 += F2;
           F_d = F*(path[1] - path[0] - mu*dt)/S/variance/dt;
           F_d2 = F_d*F_d;
           Delta += F_d;
           Delta2 += F_d2;
           tmp = 0.0;
           for (int i=1; i<=nSteps; i++)
               tmp += ((path[i]-path[i-1]-mu*dt)/Math.sqrt(variance)*(
                       (path[i]-path[i-1]-mu*dt)/variance/dt - 1.0));
           F_k = F*tmp;
           F_k2 = F_k*F_k;
           Kappa += F_k;
           Kappa2 += F_k2;
           tmp = 0.0;
           for (int i=1; i<=nSteps; i++)
               tmp += ((path[i]-path[i-1]-mu*dt)/variance);
           F_i = -nSteps*dt*F + F*tmp;
           F_i2 = F_i*F_i;
           Rho += F_i;
           Rho2 += F_i2;
        } while (M < nTrials);
        // plotting is an instance of the Plotter class
        parent.parent.plotting.before = false;

        // the uncertainties, Eq. (10.27)
        Payoff /= M;
        Payoff_error = Math.sqrt((Payoff2/M - Payoff*Payoff)/M);
        Delta /= M;
        Delta_error = Math.sqrt((Delta2/M - Delta*Delta)/M);
        Kappa /= M;
        Kappa_error = Math.sqrt((Kappa2/M - Kappa*Kappa)/M);
        Rho /= M;
        Rho_error = Math.sqrt((Rho2/M - Rho*Rho)/M);
}       // end of Mean

private void shake() {  // importance sampling
        int i, j, miss;
        double y0, y1, y2, y1_, tmp;
        double Lambda_, Lambda, W=0.0;

        miss = 0;                       // the use of miss is to tune
        for (i=1; i<=nSteps; i++) {    // the magnitude of lambda in
            miss = 0;                   // Eq. (10.32)
            y0 = path[i-1];
            y1 = path[i];
            y2 = path[i+1];
            y1_ = y1 + shift*(2.0*rand.nextDouble() - 1.0);
```

```
            tmp = (y1_ - y0 - mu*dt)*(y1_ - y0 - mu*dt);
            Lambda_ = Math.exp(-tmp/2.0/variance/dt);

            tmp = (y1 - y0 - mu*dt)*(y1 - y0 - mu*dt);
            Lambda = Math.exp(-tmp/2.0/variance/dt);

            W = Lambda_/Lambda;

            if (W >= 1.0) {                // note the global movement
               for (j=i; j<=(nSteps+1); j++) path[j] += (y1_ - y1);
//               miss += 1;
            } else {
               if (rand.nextDouble() < W) {
                  for (j=i; j<=(nSteps+1); j++) path[j] += (y1_ - y1);
//                  miss += 1;
               } else miss = 1;
            }
        }  // end of for
//        System.out.println(shift+" "+miss+" "+W);
 }      // end of method shake

 // Brownian motion of the stock price increments
 // one can equally well uses shake2()
 // however, the importance sampling technique in shake()
 // is more general, applicable to arbitrary distributions
 private void shake2() {
        for (int i=1; i<=nSteps; i++) {
            path[i] = path[i-1] + mu*dt +
                      Math.sqrt(variance*dt)*rand.nextGaussian();
        }
 }
}       // end of Path class
```

Listing 10.1 Path.java

Figures 10.2 to 10.9 show some possible stock price movements generated by the path integral Monte Carlo. Payoff of the option at maturity is calculated from the stock price at maturity from these possible stock prices.

10.7 Summary

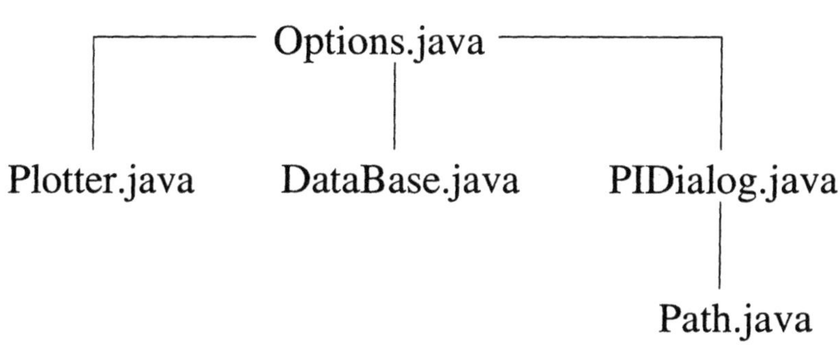

Figure 10.10. Source programs for the option pricing using path integral Monte Carlo

`Plotter.java` and `DataBase.java` for the plotting, `PIDialog.java` for user interface and `Options.java` with the `main()` method are easily obtained by modifying corresponding classes in previous chapters.

We demonstrated options pricing using path integrals. One of the advantages of the method is that it calculates, in addition to options price, options sensitivities in the same Monte Carlo path. The nature of sum over paths also makes the methodology suitable for parallel computation.

Feynman's path integral method is a powerful computational as well as theoretical tool for researchers in a variety of scientific fronts.

10.8 References and Further Reading

Feynman's path integral method was best introduced in the following treatises, R.P. Feynman and A.R. Hibbs, "Quantum Mechanics and Path Integrals", McGraw-Hill, New York (1965) and L.S. Schulman, "Techniques and Applications of Path Integration", John Wiley & Sons, Inc., New York (1981)

The Black-Scholes equation was presented in, F. Black and M.J. Scholes, "The Pricing of Options and Corporate Liabilities", *Journal of Political Economy*, **81** (1973) 637-659

A more general path-integral formula for options pricing with variable volatility was derived in, B.E. Baaquie, "A Path Integral Approach to Option Pricing with Stochastic Volatility: Some Exact Results", *J. de Phys. I (France)*, **7** (1997) 1733-1753, available at `xxx.lanl.gov/cond-mat/9708178` 22 Aug 1997

A path integral Monte Carlo evaluation of options prices can be found in, M.S. Makivic, "Numerical Pricing of Derivative Claims: Path Integral Monte Carlo Approach", *NPAC Technical Report SCCS 650*, Syracuse University, 1994

Metropolis-Hastings algorithm was shown in, W.K. Hastings, "Monte Carlo sampling methods using Markov chains and their applications", *Biometrika*, **57** (1970) 97-109

Chapter 11

DATA FITTING

An experimenter, to test the ideas or theories in mind, prepares her experiment. She excites the sample by a well controlled means. Reactions from the sample are measured by appropriate apparatus. The next step is to compare the recorded data with the theory. The task of comparison is most usually practiced by the method of *chi-square* (or least squares) fitting. Another application of chi-square fits is in data interpolation/extrapolation. Since chi-square fits are so commonly used by researchers in data analysis, we demonstrate a class which performs a 2-dimensional chi-square fit.

11.1 Chi-Square

Consider a theory which describes a phenomenon with a function, $f(x; a_1, \cdots, a_M)$. For example, $f = a_1 + a_2 \sin(a_3 x)$ models the temperature, f, of a room at time x. Parameters a_1, a_2, a_3 then describe the baseline (or mean), amplitude of fluctuation, and frequency of fluctuation, of the temperature, respectively. A series of N measurements are then made at different times in order to determine the parameters, yielding data y_i at time x_i, $i = 1, 2, \cdots, N$. Note that N has to be greater than the number of unknown parameters in the model function, which is 3 in this case. Also noted for sinusoidal signals is that the highest frequency that can be determined is limited by the the shortest time interval between successive measurements (the Nyquist critical frequency).

Measurements are subject to errors, which are inherent to either the nature of the phenomenon or the measuring device, or both. For example, in a counting measurement, the counter, having registered a number, can give a different number at the next reading. To analyze experimental data, we then seek a figure-of-merit which takes errors into account. A common figure-of-merit is the *chi-square*, defined as,

$$\chi^2(a_1, a_2, \cdots, a_M) = \sum_{i=1}^{N} \left[\frac{y_i - f(x_i; a_1, \cdots, a_M)}{\sigma_i} \right]^2, \tag{11.1}$$

where y_i and σ_i are respectively the data and its error, at measurement x_i. The task is now to minimize χ^2 by tuning the parameters, $a_1, a_2, \cdots, a_M$. The model (theory) that best fits the data is then the one whose parameter values give a minimal χ^2. When the errors are the same for all data points, σ_i can be moved out of the summation. Minimization of Eq. (11.1) is then called *least squares method.*

It is seen that terms in the sum in χ^2 are weighted: If error σ_i is large, since it appears in the denominator, the term is much discounted and contributes little to the sum. Each term can be thought of as a residual properly normalized to its measurement precision. Furthermore, if the function, with the best adjusted parameters, faithfully describes the phenomenon, the sum becomes $\chi^2 \approx \sum_{i=1}^{N} 1$ and should come up close to the number of degrees of freedom of the fit: $N - M$, where M is the number of freely adjusted parameters in the function. The usage of chi-square can therefore be two folds. Given known parameters, the value of χ^2 tests the *goodness of fit* and thus the theory (or model function). On the other hand, if the theory is well addressed, parameters specific to the sample can be extracted. We adopt the latter theme throughout the chapter. Uncertainties in the estimated parameters are equally important and will also have to be estimated.

11.2 Marquardt Recipe

The fitting function, $f(x_i; a_1, \cdots, a_M)$, can be either linear or non-linear in the parameters: $a_1, a_2, \cdots, a_M$. For non-linear chi-square minimization, analytical solutions are not available and the minimum chi-square has to be approached iteratively, starting from a set of trial parameters. First of all, we need to calculate the derivatives of χ^2 with respect to the parameters: $\nabla\chi^2(\mathbf{a})$, evaluated at the current values of the parameters $\mathbf{a}$, which is now a vector of components, $a_1, a_2, a_3, \cdots$, and so on. By definition, the negative gradient, $-\nabla\chi^2(\mathbf{a})$, gives the direction of decreasing χ^2. The new trial parameters, $\mathbf{a}_{\text{new}}$, are then chosen according to,

$$\mathbf{a}_{\text{new}} = \mathbf{a} - \text{constant}\ \nabla\chi^2(\mathbf{a}). \tag{11.2}$$

The proportional constant is deliberately chosen so as not to exhaust the downhill, the reason being that, around the true (global) minimum, there might sit local minima and that we do not want the search to be trapped in a local minimum.

On the other hand, once in the vicinity of the true minimum, the χ^2 function can be well approximated by a (M-dimensional) parabola. Moreover, a linearized $f(x_i; a_1, \cdots, a_M)$ is also an adequate approximation to itself. Under these approximations, analytical solution exists, which is very similar to Eq. (11.2), and $\mathbf{a}_{min}$ is readily calculated. D.W. Marquardt successfully and smoothly blended the two (gradient and expansion approximation) with a single steering parameter in his algorithm. The so-called Levenberg-Marquardt method has become the standard method for chi-square fits.

Even with the gradient-expansion method, it pays off to start with a good initial guess. For example, we might temporarily fix some of the better known parameters and adjust the other less certain ones during the fit. After gaining some experience in the range of the parameters, all are set free in the fit. The procedure is repeated until there is no improvement in the resulting chi-square.

11.3 Uncertainties in the Best-Fit Parameters

As important as parameter estimation are errors in the estimated parameters, $\sigma_{a_1}, \sigma_{a_2}, \cdots, \sigma_{a_M}$, which occur because of the fact that errors in the gathered data propagate to the parameters,

$$\sigma_{a_i}^2 = \sum_{i=1}^{N} \sigma_i^2 \left[\frac{\partial a_i}{\partial f(x_i; a_1, \cdots, a_M)} \right]^2 . \tag{11.3}$$

It can be shown that the diagonal elements of the inverse of the so-called curvature matrix, α_{ij},

$$\alpha_{ij} = \frac{1}{2} \frac{\partial^2 \chi^2}{\partial a_i \partial a_j}, \tag{11.4}$$

give the variances of the estimated parameters. The source program we are providing also returns these errors.

To demonstrate a fitting, we need to have data. In the next section, we show how to generate *ideal* data from a specified distribution by Monte Carlo. Each data point is also associated with its error. We then show that *true* values of the parameters are indeed within the range bounded by the best-fit estimates plus/minus their uncertainties returned from the chi-square fit.

11.4 Arbitrary Distributions by Monte Carlo

Monte Carlo techniques are widely used to better understand the systematics of a complex experiment. They are also used, for example, to perform multi-dimensional integration. In this section, Monte Carlo method is demonstrated to generate arbitrary distribution functions. If particle interactions with various materials are cast into distributions, then the effect of, for instance, gamma-rays passing through human tissues can be simulated.

Java language comes with the `java.lang.Math.random()` method, which, when called successively, generates a series of numbers randomly distributed in the half open interval of $[0, 1)$. The period of the series is long enough that the numbers are in practice statistically independent. From this uniform random number generator ζ, uniform distributions of another range, $[a, b)$, can be generated,

$$a + (b - a)\zeta. \tag{11.5}$$

To generate a non-uniform (normalized) distribution, $f(x)$, from `random()`, we use the Acceptance-Rejection (Von Neumann) method of Section 7.3. Suppose the range of x is from a to b, and the maximum of f is $f_{\max}$. The method then reads,

1. draw a random number r_{try} between a and b
2. evaluate $f(r_{\text{try}})$
3. draw a random number r between 0 and $f_{\max}$
4. if $r \leq f(r_{\text{try}})$ accept r_{try}, otherwise go to step 1

It is seen from the method that pairs of numbers, (r_{try}, r), are randomly drawn from the rectangle, $(a, 0), (b, 0), (b, f_{\max}), (a, f_{\max})$. Step 4 accepts those under the function curve. The selected samples thus follow the desired distribution of $f(x)$. Extension to higher dimensions is straightforward: r_{try} is now a vector of the appropriate dimensions. Points in the hyper-volume are randomly drawn and the accepted are those under the hyper-surface defined by the distribution function. It is noted that if the function to be sampled peaks in some narrow region, the above procedure wastes lots of time (random number generators are typically slow) probing unimportant space. It is suggested in this case that the distribution function be split into several parts, each having its maximum. Procedures are then repeated in each part. Wasteful drawings can then be greatly reduced. This is another example of importance sampling.

We present a class which samples from the following 2-dimensional distribution (in x and $\cos\theta$),

$$\begin{aligned} f(x, \cos\theta) = \; & (x^2 - x_0^2)^{\frac{1}{2}} \Big\{ 6x(1 - x) \\ & + \frac{4}{3}\rho(4x^2 - 3x - x_0^2) \\ & + 6\eta x_0(1 - x) \\ & + P_\mu \xi \cos\theta (x^2 - x_0^2)^{\frac{1}{2}} \left[2(1 - x) \right. \\ & \left. + \frac{4}{3}\delta(4x - 4 + (1 - x_0^2)^{\frac{1}{2}})\right] \Big\}, \end{aligned} \tag{11.6}$$

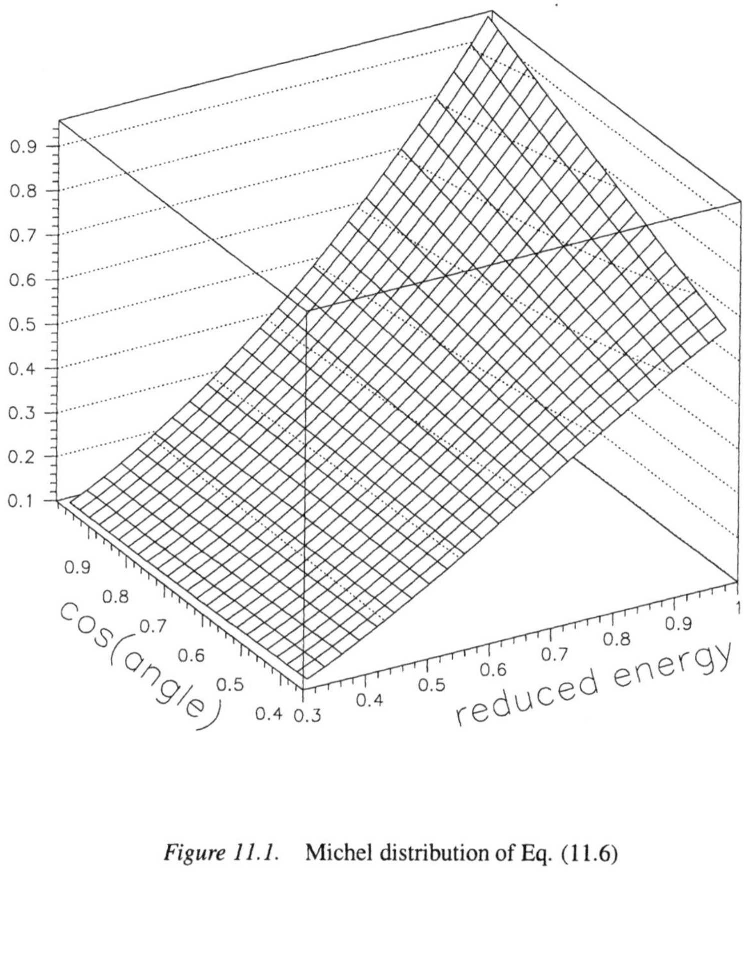

Figure 11.1. Michel distribution of Eq. (11.6)

where $x_0 < x < 1$, $0 < \theta < 90$ degrees, and x_0 is a constant. Furthermore, $\rho = 0.75$, $\eta = 0$, $P_\mu \xi = 1$, and $\delta = 0.75$. Equation (11.6) is the theoretical distribution of the energies and angles of positrons (negative electrons) emerged from muon-decays at rest. It is called Michel distribution and is plotted in Figure 11.1.

Method `Michel(double x, double y)` in class `Michel` (Listing 11.1) returns the function value at input arguments x and $y = \cos\theta$. In the variable field of the class are defined some constants which are then used in the class constructor to calculate x_0. The range and number of bins in each dimension of the histogram are specified in the constructor. We specify a narrower range than theory (kinematics) allows because of limitations on the detector capability and/or experimental geometry. Since class `Michel` inherits `Thread`, we have to supply our `run()` method, which is invoked when the thread `starts`. In the present case, the method wrapped inside `run()` is `initialize()` which samples the Michel distribution of Eq. (11.6) and populates the histogram in Figure 11.2. The `do-while` loop codes the acceptance-rejection algorithm. The rest in method `initialize()` deals with re-arranging the accumulated 2-dimensional histogram, at points (x, y), into a one-dimensional array, which is to be passed to the `SurfaceFit` class for chi-square fits. Also noted is that the distribution is normalized.

```
/*      Sun-Chong Wang
        TRIUMF
        4004 Wesbrook Mall
        Vancouver, V6T 2A3
        Canada
        e-mail: wangsc@triumf.ca

        Michel.java samples the Michel distribution Eq. (11.6)
        and generates a histogram */

import java.lang.*;
import java.util.*;

public class Michel extends Thread {

 final double mass_e = 0.510998902;
 final double mass_u = 105.6583568;
 final double sigma_xy = 100.e-6; // resolution of the
                                  // detector in meters
 double norm;
 int deduction;                   // over- or under-flow
 double x_0, Wmue;
 int ndata;
 double[][] x;
 double[] y;

 double minx, maxx, miny, maxy;
 int nx, ny;
 long dataVol;

 Random rand;

 public Michel (long N, long seed) {
        rand = new Random();
        rand.setSeed(seed);

        Wmue  = (mass_e*mass_e+mass_u*mass_u)/2.0/mass_u;
        x_0   = mass_e/Wmue;

        dataVol = N;
        minx = 0.3;  // energy in unit of Wmue
        maxx = 1.0;
        miny = 0.342020143;  // cos(theta), theta = 70 degree
        maxy = 0.984807753;  // cos(theta), theta = 10 degree

        nx    = 370; // # of bins in energy
        ny    = 60;  // # of bins in cos(theta)

        ndata = nx*ny;
        x     = new double[ndata][2];
        y     = new double[ndata];
 }

 public void run() {
```

```
        initialize();
    }

    public void initialize() {
        double binwx, binwy, e, theta;
        double p, max, tmp;
        int n, ix, iy;

        binwx = (maxx-minx)/nx;
        binwy = (maxy-miny)/ny;

        max = 0.0;
        norm = 0.0;
        for (int i=0; i<nx; i++) {
            for (int j=0; j<ny; j++) {
                e     = minx + binwx*(i+0.5);
                theta = miny + binwy*(j+0.5);
                n = i*ny + j;
                x[n][0] = e;
                x[n][1] = theta;
                tmp = Michel(e, theta);
                if (tmp >= 0.0) norm += tmp;
                else System.out.println("Error in Michel spectrum");
                if (tmp > max) max = tmp;
            }
        }

        deduction = 0;
        for (int i=0; i<dataVol; i++) {
            do {    // sampling by acceptance-rejection method
                e = minx + (maxx-minx)*rand.nextDouble();
                theta = miny + (maxy-miny)*rand.nextDouble();
                p = Michel(e, theta)/max;
            } while (rand.nextDouble() >= p);

            if (e > 1.0 || e < minx || theta > maxy || theta < miny) {
                deduction += 1;
            } else {
                ix = (int) ((e-minx)/binwx);
                iy = (int) ((theta-miny)/binwy);
                n = ix*ny + iy;
                y[n] += 1.0;                      // histogramming
            }
        }
        System.out.println("overflow = "+deduction);
    }

    public double Michel(double x, double y) {    // Eq. (11.6)
        if (x > 1.0 || x < x_0 || y > 1.0 || y < 0.0) return 0.0;
        else return Math.sqrt(x*x-x_0*x_0)*(6.0*x*(1.0-x)+
              (4.0*x*x-3.0*x-x_0*x_0)+y*Math.sqrt(x*x-x_0*x_0)*
              (2.0*(1.0-x)+(4.0*x-4.0+Math.sqrt(1.0-x_0*x_0))));
    }
}       // end of class
```

Listing 11.1 Michel.java

11.5 A Surface Fit Example

In the last section, Michel distribution data (i.e., the histogram) were generated with known values of the Michel parameters, $\rho = 0.75, \eta = 0, P_\mu\xi = 1$, and $\delta = 0.75$, in the Monte Carlo. Each bin of the histogram contains counts whose uncertainty is, according to Poisson statistics, equal to the square root of the count. The example in this section is to fit the data to the Michel distri-

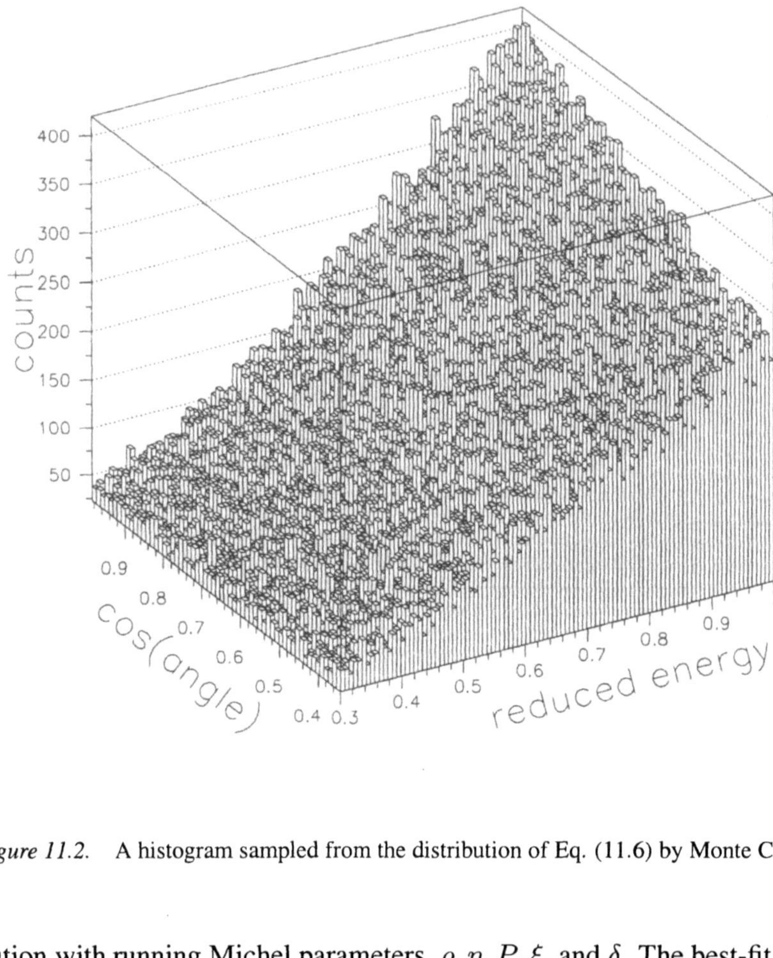

Figure 11.2. A histogram sampled from the distribution of Eq. (11.6) by Monte Carlo

bution with running Michel parameters, $\rho, \eta, P_\mu\xi$, and δ. The best-fit values of $\rho, \eta, P_\mu\xi$, and δ, together with their uncertainties, are to be shown to enclose the *true* values of 0.75, 0.0, 1.0, and 0.75.

Listing 11.2 is the chi-square fit class implementing the Marquardt method. To perform your custom chi-square fitting, you need to provide your data and replace the fitting function `func()` with your own. Derivatives with respect to fitting parameters are calculated numerically by the method `dfunc_da()`. In the beginning of the class is defined the number of parameters to fit. It is 5 in the present case: four Michel parameters plus a scaling (normalization) constant.

```
/*      Sun-Chong Wang
        TRIUMF
        4004 Wesbrook Mall
        Vancouver, V6T 2A3
        Canada
        e-mail: wangsc@triumf.ca
```

```
        SurfaceFit.java does chi-square fitting
        with the Marquardt method */

import java.lang.*;
import java.util.*;

public class SurfaceFit {

 final int ma = 4; // # of free parameters

 double x_0;
 int nx, ny;
 int ndata;
 double[][] x;
 double[] y, sig;
 double[] weight;
 double[] a;
 double[] dyda;
 double[] delta_a;
 double[] sigmaa;

 double chisqr;
 double flamda;
 long dataVol;
 double a4;  // the normalization constant

 Michel michel1, michel2;
 Random rand;

 public static void main(String args[]) {
        SurfaceFit lm = new SurfaceFit();

        try {
            lm.michel1.start(); // two threads
            lm.michel2.start();
            lm.michel1.join();
            lm.michel2.join();
        } catch (InterruptedException e) {}

        lm.merge();
        lm.Go();
 }

 public SurfaceFit() {
        super();
        rand    = new Random();
        dataVol = 1000000000; // total number of counts

        michel1 = new Michel(dataVol/2, rand.nextLong());
        michel2 = new Michel(dataVol/2, rand.nextLong());

        x_0     = michel1.x_0;

        nx      = michel1.nx;
        ny      = michel2.ny;

        ndata   = nx*ny;
        x       = michel1.x;
        y       = new double[ndata];
        sig     = new double[ndata];
        weight  = new double[ndata];
        a       = new double[ma];
        sigmaa  = new double[ma];
        dyda    = new double[ma];
        delta_a = new double[ma];

        chisqr  = 0.0;
```

```
        flamda  = -1.0;

        a[0] = 0.75; // initial guess
        a[1] = 0.0;
        a[2] = 1.0;
        a[3] = 0.75;
}

private void merge() {
        for (int i=0; i<ndata; i++) {
            y[i] = michel1.y[i]+michel2.y[i];
            if (y[i] == 0.0) sig[i] = Double.MAX_VALUE;
            else sig[i] = Math.sqrt(y[i]);
        }
        a4 = (dataVol-michel1.deduction-michel2.deduction)/michel1.norm;
}

public void Go() {
        int j=0, imax=6;
        double oldc;

        flamda = -1.0;
        Marquardt();
        oldc = chisqr;
        System.out.println(chisqr+" at iteration = 0");
        do {
            oldc = chisqr;
            Marquardt();
            j += 1;
            System.out.println(chisqr+" at iteration = "+j+" "+flamda);
        } while (((oldc-chisqr)/chisqr) > 0.001 && j < imax);

        flamda = 0.0;
        Marquardt();
        for (j=0; j<ma; j++) {
            System.out.println(a[j]+" +- "+sigmaa[j]);
        }
        System.out.println(a4);
        System.out.println("chi-sq = "+chisqr);
        System.out.println("NDF = "+ndata+" - "+ma);
}

private void dfunc_da(double[] z, double[] a, double[] da) {
// calculate derivatives numerically
        double tmp;

        for (int i=0; i<ma; i++) {
            if (da[i] != 0.0) {
               tmp = a[i];
               a[i] += da[i];
               dyda[i] = func(z,a);
               a[i] -= (2.0*da[i]);
               dyda[i] -= func(z,a);
               dyda[i] /= (2.0*da[i]);
               a[i] = tmp;
            } else
               dyda[i] = 0.0;
        }
}

private double func(double[] z, double[] a) {
        double tmp, x, y;

        x = z[0];
        y = z[1];
        // Michel distribution, Eq. (11.6)
```

```
        if (x <= 1.0 && x >= x_0) {
           tmp = Math.sqrt(x*x-x_0*x_0)*(6.0*x*(1.0-x)+
                 4.0/3.0*a[0]*(4.0*x*x-3.0*x-x_0*x_0)+
                 6.0*a[1]*x_0*(1.0-x)+
                 a[2]*y*Math.sqrt(x*x-x_0*x_0)*(2.0*(1.0-x)+
                 4.0/3.0*a[3]*(4.0*x-4.0+Math.sqrt(1.0-x_0*x_0))));
           return a4*tmp;
        } else {
           return 0.0;
        }
}

private void Marquardt() {
        int i,j,k;
        double dy, chisq1;
        double[][] alpha = new double[ma][ma];
        double[] beta = new double[ma];
        double[] b = new double[ma];
        Matrix array = new Matrix(ma,ma);
        boolean gradient;

        if (ndata <= ma) {
           chisqr = 0.0;
           System.out.println("Not enough data points");
           return;
        }

        for (j=0; j<ma; j++) {
            beta[j] = 0.0;
            for (k=0; k<=j; k++) alpha[j][k] = 0.0;
        }

        if (flamda < 0.0) {
           flamda = 0.001;
           for (i=0; i<ndata; i++) weight[i] = 1.0/sig[i]/sig[i];
           for (j=0; j<ma; j++) {
               if (a[j] == 0.0) delta_a[j] = 0.01;
               else delta_a[j] = a[j]/100.0;
           }
        }

        chisq1 = 0.0;
        for (i=0; i<ndata; i++) {
            dy = y[i] - func(x[i],a);
            dfunc_da(x[i],a,delta_a);
            for (j=0; j<ma; j++) {
                beta[j] += (weight[i]*dy*dyda[j]);
                for (k=0; k<=j; k++)
                    alpha[j][k] += (dyda[j]*dyda[k]*weight[i]);
            }
            chisq1 += dy*dy*weight[i];
        }
        chisq1 /= (ndata-ma); // reduced chi-square

        for (j=0; j<ma; j++)
            for (k=(j+1); k<ma; k++) alpha[j][k] = alpha[k][j];

        do {
           for (j=0; j<ma; j++) {
               for (k=0; k<ma; k++) array.M[j][k] = alpha[j][k];
               array.M[j][j] = alpha[j][j]*(1.0+flamda);
           }
           try {
               array.inverse();
           } catch (MyMatrixExceptions mme) {
```

```
                System.out.println(mme.getMessage());
            }

            if (flamda != 0.0) {
               for (j=0; j<ma; j++) {
                   b[j] = a[j];
                   for (k=0; k<ma; k++) b[j] += (beta[k]*array.M[j][k]);
               }

               chisqr = 0.0;
               for (i=0; i<ndata; i++) {
                   dy = y[i] - func(x[i],b);
                   chisqr += dy*dy*weight[i];
               }
               chisqr /= (ndata-ma);

               if (chisqr > chisq1) { // more gradient
                  flamda *= 10.0;
                  gradient = true;
               } else {               // more expansion
                  flamda /= 10.0;
                  gradient = false;
               }
            } else gradient = false;
         } while (gradient == true);

         for (j=0; j<ma; j++) {
             if (flamda == 0.0) {
                sigmaa[j] = Math.sqrt(array.M[j][j]);
             } else {
                delta_a[j] = b[j]-a[j];
                a[j] = b[j];
             }
         }
   }     // end of method
}        // end of class
```

Listing 11.2 SurfaceFit.java

The `main()` method instantiates an object of class `SurfaceFit` and then prepares the high statistics data, which are accomplished by two threads. Separate histograms from each thread are then combined by the method `merge()` before fitting. The fitting method is `Marquardt()`, which is based on the Fortran routine in *Data Reduction and Error Analysis for the Physical Sciences* by P.R. Bevington. Transition between the gradient and expansion method in the Marquardt recipe is controlled by a parameter called `flamda`. It's initially set to a negative number and the initial chi-square value is calculated. Then `flamda` will get smaller (divided by 10.0) if the fit improves. A smaller `flamda` dictates the algorithm toward the expansion method. On the other hand, if the new set of trial parameters increases the chi-square, `flamda` is multiplied by 10.0, shifting the method toward the gradient (steepest descent) method. When the chi-square does not improve significantly any more, `flamda` is set to zero, directing the recipe to calculate the covariance matrix before quitting. Method `Go()` codes the above procedure to call `Marquardt()`.

We fitted the Monte Carlo data to the Michel distribution function with the four freely adjusting Michel parameters. The fifth parameter, the normalization constant, is in fact the total number of counts generated. Its value was known and has thus been kept fixed. The result of the fit is shown in Table 11.1. It's seen that the best-fit values plus/minus their uncertainties (statistical errors) enclose the true values.

Table 11.1. 10^9 pairs of energy and angle are generated by Monte Carlo according to the distribution of Eq. (11.6) given $\rho = 0.75, \eta = 0, P_\mu\xi = 1$, and $\delta = 0.75$. The resulting 2-dimensional histogram is fitted to Eq. (11.6) with now freely adjusting parameters (all free simultaneously). The 2nd column are the re-constructed parameters. The ratio of the chi-square to the number of degrees of freedom is 1.003. The last column are the estimated errors in the best-fit parameters.

parameters	best-fit value	statistical error
ρ	0.7499	3×10^{-4}
η	0.0020	57×10^{-4}
$P_\mu\xi$	0.9997	4×10^{-4}
δ	0.7504	4×10^{-4}

11.6 Summary

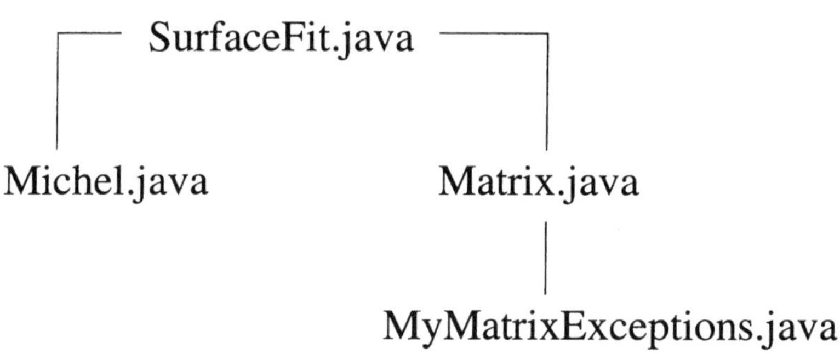

Figure 11.3. Source programs for the Marquardt chi-square fit

`Matrix.java` and its exceptions class are the same ones as in Chapter 1.

We demonstrated a 2-dimensional chi-square fit by the use of the Levenberg-Marquardt method. The user can supply her fitting function for tailored use. Derivatives of the function with respect to the parameters are evaluated numerically.

Monte Carlo techniques were applied to generate a specific distribution, which was used here to provide an ideal set of data.

Chi-square fits are widely used to estimate parameters from measured data. For fitting functions non-linear in parameters, iterative procedures are used to search the parameters that minimize the chi-square. Levenberg-Marquardt method cleverly integrates the gradient (scaled steepest descent) method, which avoids pitfalls of local minima but is inefficient, with the expansion method, which leads to the minimum efficiently in the vicinity of a valley.

11.7 References and Further Reading

The Levenberg-Marquardt method was suggested by Levenberg and put forth by Marquardt in, D.W. Marquardt, "An Algorithm for Least-Squares Estimation of Nonlinear Parameters", *J. Soc. Ind. Appl. Math.*, **11**, (1963) 431-441

A textbook on statistical data analysis is, P.R. Bevington, "Data Reduction and Error Analysis for the Physical Sciences", McGraw-Hill Book Co., New York (1969)

A similar Marquardt method in C can be found in section 14.4 of, W.H. Press, B.P. Flannery, S.A. Teukolsky, and W.T. Vetterling, "Numerical Recipes in C", Cambridge University Press, New York (1988)

Chapter 12

BAYESIAN ANALYSIS

With Bayesian techniques, programs can be written to select, among possible choices, the model which best explains the data. The program therefore makes inferences. Applications of Bayesian inference can be seen in pattern/image recognition, radar target identification, medical diagnosis, relevant-gene scoring, text/e-mail classification, and so on. Inclusion of Bayes theorem in neural networks also enhances artificial intelligence's capability in such a way that it deals better with a real world full of errors and uncertainties.

12.1 Bayes Theorem

The logic used for model selection is based on probability theory. We write down the joint probability of propositions H, D, and I as,

$$\begin{aligned} P(H, D, I) &= P(H|D, I)\, P(D|I)\, P(I) \\ &= P(D|H, I)\, P(H|I)\, P(I). \end{aligned} \tag{12.1}$$

Equating the factorizations readily gives Bayes theorem,

$$\begin{aligned} P(H|D, I) &= \frac{P(D|H, I)\, P(H|I)}{P(D|I)} \\ &\propto P(D|H, I)\, P(H|I). \end{aligned} \tag{12.2}$$

Since I appears in every term in the equation as a given condition, it can be thought of as the knowledge or information common to the analysis under consideration. It is sometimes dropped in order not to clutter the formula, but keep in mind that a common ground exists based on which discussions are made. Now suppose H stands for a certain hypothesis and D for data. Equation (12.2) then relates the probability of the hypothesis given the data, to the product of the probability of data given the hypothesis and the probability of the hypothesis. Furthermore, if we have a probability distribution, $P(H_i)$, $i = 1, 2, \cdots, n$, for the various hypotheses H_i, $i = 1, 2, \cdots, n$, then a set of Eq. (12.2) with H_i's can be written down. After all products of the probabilities, $P(D|H_i|I)\,P(H_i|I)$, $i = 1, 2, \cdots, n$ are evaluated, the right, or in statistical terms, the most probable, hypothesis, is the one which gives a maximum $P(H_i|D, I)$. For example, imagine H_1 corresponds to the hypothesis of a certain character, H_2 to that of the other, and the like. The right character to be picked up is the one which gives the greatest $P(H|D, I)$. This is how inferences are made based on probability in *optical character recognition* software.

With the hypothesis H, $P(D|H, I)$ quantifies the chance of getting the data. It is called likelihood function, and $P(H|I)$ is called *prior* probability of H. $P(H|D, I)$ is the *posterior* probability. Hypothesis (or model, function) H normally contains parameters, just like the Michel distribution function of last chapter has four Michel parameters in it.[1] The method of parameter estimation by maximizing $P(H|D, I)$ is called MAP, for maximum *a posteriori*. Another way of parameter estimation is by calculating the mean of the parameter(s) with the posterior probability distribution. This way, we can also estimate the uncertainty in the mean.

12.2 Principle of Maximum Entropy

The next step after Bayes theorem is to assign probabilities. The principle of maximum entropy provides an objective way of achieving it. Entropy, S, a quantity central in thermodynamics or information theory, is a measure of randomness or ignorance of a system, subject to some constraints. Its simplest definition is,

$$S = -\sum_i P_i \log P_i, \tag{12.3}$$

where P_i is the probability of state (event) i. The principle of maximum entropy states that the state of a system, in equilibrium with the rest of the world, is the one that has a maximum entropy.

Let's see an application of the principle of maximum entropy in a simple example. When tossing a dice, we want to assess the probability of getting one of the six numbers. Assume that we are ignorant of any defect which would lead to a preference for one of the faces. The entropy of the dice is written down as in Eq. (12.3), with P_i, $i = 1, 2, \cdots, 6$ for the probability of showing up number $1, 2, \cdots, 6$, respectively. The task is now to maximize the entropy

[1] Michel distribution models the energy and angular distribution of the outgoing positions from muon decays at rest. Precise measurement of the Michel parameters can test the Standard model of particle physics, which is an encompassing model of modern physics to describe what constitute material and how they interact with each other.

subject to the constraint,

$$\sum_{i=1}^{6} P_i = 1. \tag{12.4}$$

This normalization condition merely asserts that the face that turns up after a toss must be one of the six faces. Since $(\sum_{i=1}^{6} P_i - 1)$ is equal to zero, it makes no difference if we multiply it by some number α, which is called *Lagrangian multiplier*, and then subtract the result (zero) from S. Equation (12.3) now becomes,

$$S = -\sum_{i=1}^{6} P_i \log P_i - \alpha\big(\sum_{i=1}^{6} P_i - 1\big). \tag{12.5}$$

It's seen that when we search for the P_i's which maximize the S in Eq. (12.5), the value of S will be dragged down if the sum of all the P_i's deviates from zero. The Lagrangian term therefore acts as a penalty against violations of the normalization condition. Take the derivative of S with respect to each P_i and to α and then set each derivative equal to zero. The 7 unknowns, P_i, $i = 1, 2, \cdots, 6$ and α are then solved. The result is $P_i = \frac{1}{6}$, for $i = 1, 2, \cdots, 6$, as one would expect, and $\alpha = 1 - \log 6$.

Likewise, if we know the mean and standard deviation of the distribution of some variable, ϵ, its probability density, $P(\epsilon)$, can be assigned by the principle of maximum entropy. This time, we have, apart from the normalization, two more Lagrangian terms corresponding to the following two constraints,

$$\bar{\epsilon} = \int d\epsilon\, \epsilon P(\epsilon) = 0, \tag{12.6}$$

and

$$\sigma = \left[\int d\epsilon\, (e - \bar{\epsilon})^2 P(\epsilon)\right]^{1/2}. \tag{12.7}$$

The mean of the distribution is assumed to be zero in Eq. (12.6), and the standard deviation is given in Eq. (12.7). Again, following the canonical procedure of finding extrema, we arrive at the familiar form,

$$P(\epsilon) = \frac{1}{\sqrt{2\pi}\sigma} \exp\Big[-\frac{1}{2}\big(\frac{\epsilon}{\sigma}\big)^2\Big]. \tag{12.8}$$

The exponent of the above Gaussian (or normal) distribution reminds us of the sum of normalized residuals (chi-square) in last chapter.

12.3 Likelihood Function

In a measurement of a signal, s, the acquired data, d, is usually accompanied by noise, ϵ, introduced inevitably from within the measuring equipment. This is expressed as,

$$d = s + \epsilon. \tag{12.9}$$

Instrumental (for example, electronics) noise can go up or down with an average amplitude of zero. The standard deviation of the noise amplitude is a property of the instrument, and is usually specified in the product's specs sheet. Errors can also be statistical in nature as in counting experiments. Re-arranging Eq. (12.9), we get the residual, $d - s = \epsilon$. The chance of getting the data, d, assuming the signal, s, is therefore, by virtue of Eq. (12.8),

$$P(d|s) = P(\epsilon) = \frac{1}{\sqrt{2\pi}\sigma} \exp\Big[-\frac{1}{2}(\frac{d-s}{\sigma})^2\Big]. \tag{12.10}$$

In a series of n measurements for the signal s, $S = s_1, s_2, \cdots, s_n$, we record a set of n data, $D = d_1, d_2, \cdots, d_n$. If the measurements are independent, the joint probability of errors can be factored and the probability $P(D|S)$ is written down immediately,

$$\begin{aligned} P(D|S) &= P(d_1, d_2, \cdots, d_n | s_1, s_2, \cdots, s_n) \\ &= P(\epsilon_1, \epsilon_2, \cdots, \epsilon_n) = P(\epsilon_1)P(\epsilon_2)\cdots P(\epsilon_n) \\ &= \prod_{i=1}^{n} \frac{1}{\sqrt{2\pi}\sigma_i} \exp\Big[-\frac{1}{2}(\frac{d_i - s_i}{\sigma_i})^2\Big] \\ &\propto \exp\Big[-\frac{1}{2}\sum_{i=1}^{n}(\frac{d_i - s_i}{\sigma_i})^2\Big]. \end{aligned} \tag{12.11}$$

Recall that, in Chapter 11, we often anticipate signals of some pattern. That is, s can be described by some function, $s = s(a, b, c, \cdots)$ with parameters, $a, b, c, \cdots$. The parameters can, for instance, be temperature coefficient, particle momentum, and so on, and are of interest to us. Their values are estimated by tuning them in $s(a, b, c, \cdots)$ so that the likelihood of producing the set of data, $P(D|S(a, b, c, \cdots))$, is maximal. Maximizing $P(D|S)$ of Eq. (12.11) is tantamount to minimizing the sum in the exponent, yielding the familiar chi-square minimization,

$$\chi^2(a, b, c, \cdots) = \sum_{i=i}^{n} \Big[\frac{d_i - s_i(a, b, c, \cdots)}{\sigma_i}\Big]^2. \tag{12.12}$$

The approach justifies the choice of chi-square as the figure-of-merit for data modeling in last chapter. We proceed to demonstrate, beyond parameter estimation, an example of Bayes analysis.

12.4 Image/Spectrum Restoration

Most often signals are blurred because of limitations of the instrument. It happens in images taken by telescopes (magnetic resonance imaging) in as-

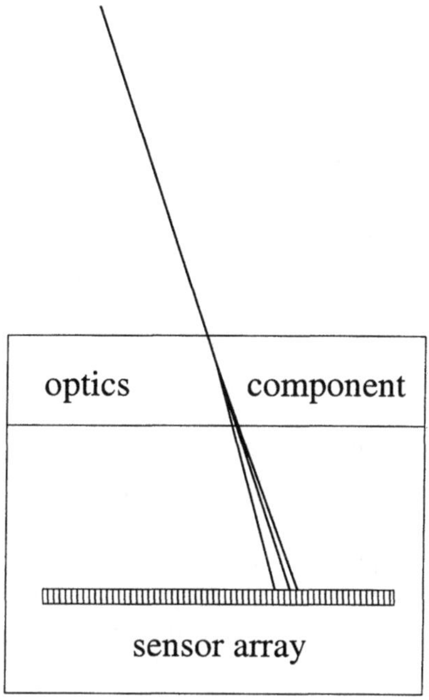

Figure 12.1. Effect of imperfect focusing: an otherwise point object looks extended

tronomy (medical diagnosis) or in spectra by photodetectors in various analysis labs. Figure 12.1 illustrates such a smearing. The blurring, together with the noise, can be expressed as,

$$d(x) = \int dx'\, r(x - x')s(x') + \epsilon(x), \tag{12.13}$$

or, in a discrete form,

$$d_i = \sum_j r_{ij}s_j + \epsilon_i, \tag{12.14}$$

where $r(x - x')$ is the so-called response function, giving the output of the detector at element x with an input to element x'. If $r(x - x')$ were a delta function, then $d(x) = s(x) + \epsilon(x)$ and there would be no blurring. Normally the response function is represented by a Gaussian function whose standard deviation quantifies extent of blurring. This standard deviation then sets the resolution of the instrument, another property in the specs sheet.

Solving Eq. (12.13) for $s(x)$ is an *inverse problem* in mathematics. If we unfold the signal by simply divide Eq. (12.13) by $r(x - x')$, a serious problem might occur in that noise gets amplified for small values of $r(x - x')$. Instead, we treat the deconvolution (unfolding) problem by applying Bayesian analysis.

By Bayes theorem, the probability, $P(S, R|D)$, of the signal, S, and the blurring, R, given the set of data, D, is,

$$P(S, R|D) \propto P(D|S, R)\, P(S, R) = P(D|S, R)\, P(S|R)\, P(R). \quad (12.15)$$

If we don't have any bias against any particular response (or resolution) function, $P(R)$ is the same for different R and therefore $P(R)$ is a constant. Equation (12.15) then becomes,

$$P(S, R|D) \propto P(D|S, R)\, P(S|R). \quad (12.16)$$

The elemental sensing unit of a camera registers the light originating from part of an extended object in the sky. The output of the sensor unit goes to a pixel of the image. Its content is proportional to the brightness of the light. A pixel of an image is equivalent to a *bin* in a histogram in spectroscopic measurement. Suppose the effect of the aperture of an optics (or the particle focusing component of a detector) is to *evenly* spread the light, heading to pixel i, across the adjacent pixels, $i - 1, i, i + 1$. The response function is then,

$$r_{ij} = \begin{cases} \frac{1}{3}, & j = i - 1, i, i + 1 \\ 0, & \text{otherwise} \end{cases} \quad (12.17)$$

In this simplified case, the degree of smearing is managed by the number of non-vanishing element in r_{ij}. Therefore, to increase smearing, we could simply change the response to, $r_{ij} = 1/5$ for $j = i - 2, i - 1, i, i + 1, i + 2$. A note in passing: when it is generalized to a 2-dimensional case, isotropy requires that none-zero elements of the response function be enclosed in circles if the aperture design is symmetrical.

The response function of Eq. (12.17) suggests that pixels $i - 1$, i, and $i + 1$ are no longer independent. In other words, the three separate pixels can be replaced by a single big one. In the language of curve (chi-square) fitting, the number of degrees of freedom is less than the total number of pixels (data points). The prior probability $P(S|R)$ would thus favor a proper set of pixels in which the pixel sizes vary, reflecting the smearing of Eq. (12.17). In other words, counting statistics argues that the prior goes like,

$$P(S|R) = \frac{N!}{n^N \prod_{i=1}^{N} N_i!}, \quad (12.18)$$

where n and N are, respectively, the total number of pixels and counts, and N_i is the number of counts at pixel i. To maximize the posterior probability $P(S, R|D)$, we can decrease the total number of pixels n in Eq. (12.18) while maintaining an adequate likelihood function $P(D|S,R)$. The deconvolution problem then becomes the one to find the proper pixel base. This approach is called Pixon algorithm.

Suppose the response is such that there can be either no smearing or the smearing of the form of Eq. (12.17). For example, if the range of smearing of a pixel is either 0 or 1 and N pixels form an image, there are then 2^N possible sets of pixel bases. If the aperture spreads hits further to the next neighboring pixels, the number of choices grows rapidly to an astronomical number, 3^N. Together with the parameters, $a, b, c, \cdots$ in the signal function, $S(a,b,c,\cdots)$, image restoration by maximizing $P(S,R|D)$ is challenging. A solution is introduced in the following section.

12.5 An Iterative Procedure

Since both S and R contain variables to optimize, we treat them alternatively in an iterative loop:

1. maximize $P(S,R_0|D) \propto P(D|S,R_0)P(S,R_0)$ given R_0
2. maximize $P(S_0,R|D) \propto P(D|S_0,R)P(S_0,R)$ given S_0
3. quit if converge, otherwise go to step 1.

The loop firstly optimizes $P(S,R_0|D)$ by tuning the parameters $a, b, c, \cdots$ in $S(a,b,c,\cdots)$ with a default pixel base of no-smearing. This step of optimization is the standard method of parameter estimation by minimizing the chi-square introduced in last chapter. Step 2 in the loop is the optimization by trying possible combinations of pixel sizes. Since the configuration space is huge, the technique of simulated annealing of Chapter 4 is used. The objective function of the annealing is,

$$\chi^2(\beta's) = \sum_{i=1}^{N}\left[\frac{d_i - \sum_j r_{ij}(\beta)s_j(a_0,b_0,c_0,\cdots)}{\sigma_i}\right]^2, \tag{12.19}$$

where the summation in j is over nonvanishing elements of $r_{ij}(\beta)$, and the β of pixel i can be either 0 or 1 (or even 2):

$$\begin{aligned}
r_{ij}(0) &= \begin{cases} 1, & j=i \\ 0, & \text{otherwise} \end{cases} \\
r_{ij}(1) &= \begin{cases} \frac{1}{3}, & j=i-1,i,i+1 \\ 0, & \text{otherwise} \end{cases} \\
r_{ij}(2) &= \begin{cases} \frac{1}{5}, & j=i-2,i-1,i,i+1,i+2 \\ 0, & \text{otherwise} \end{cases}
\end{aligned} \tag{12.20}$$

A heuristic 'cooling' procedure is to increase the β of each pixel slowly from 0 to 2 (or higher depending on the maximum smearing conceived). When going back to step 1 in the loop, the chi-square function in the Levenberg-Marquardt method is again Eq. (12.19), with now free parameters $a, b, c, \cdots$ and the fixed base β's: $\chi^2(a, b, c, \cdots)$.

12.6 A Pixon Example

We provide an implementation of the above iterative deconvolution. The procedure in the loop requires two tasks which are undertaken by two classes `LMFit` and `Anneal` introduced earlier in the book (`LMFit`, being a one dimensional chi-square fit, is a simple version of `SurfaceFit` of last chapter). The class `Pixon`, Listing 12.1, contains the `main()` method, which simply creates an instance of `Pixon`, whose constructor prepares for the (ideal) histogram data: the Michel spectrum of last chapter. In the present illustration, however, we consider deconvoluting 1-dimensional histograms. The 2-dimensional Michel distribution (one in energy, the other in angle) is now sampled at a fixed angle, which is chosen to be 35 degree in present illustration.

It can be shown that errors, σ, in spatial measurement of the detector incur an uncertainty in the particle momentum p ($\approx$ energy in the energy range of interest) in the following way,

$$\sigma_p = \left[\left(\frac{\kappa q B}{\sin\theta} \right)^2 \frac{1}{2N} + \frac{p^2}{(\Delta z)^2} \left(\frac{\cos^3\theta}{\sin\theta} \right)^2 \frac{6}{N(N+1)(2N+1)} \right]^{\frac{1}{2}} \sigma, \quad (12.21)$$

where B is the magnetic field strength, κq is some constant, N the number of x (or y) coordinate measurements and Δz the spacing between the measuring planes. We therefore deliberately smear the energy sampled from the theoretical (ideal) Michel distribution [Eq. (11.6) and Figure 11.1 at angle = 35 degree] by the above uncertainty before it is binned into histogram,

$$E \to E + G(0, \sigma_E), \quad (12.22)$$

where $G(0, \sigma_E)$ stands for a Gaussian distribution with mean zero and standard deviation σ_E which is a function of E (and also angle θ). Again, $E \approx p$ and thus $\sigma_E \approx \sigma_p$ of Eq. (12.21) in the energy range of our interest. This is done in class `Michel`. Note that two threads are created and work in parallel to generate the Michel spectrum which will contain a total of 10^9 data points. Accumulation of this high statistics spectrum takes most of the CPU time, and justifies the concurrent use of a dual CPU's by two threads. In `Pixon`, the `initialize()` method defines the binning of the histogram. The method `merge()` combines the histogram generated by each thread, in which random number generator is seeded differently in order to ensure statistically independent data. Method `Go()` codes the iterative procedure of alternative calls to the Levenberg-Marquardt curve fitting and the simulated annealing method.

```
/*      Sun-Chong Wang
        TRIUMF
        4004 Wesbrook Mall
        Vancouver, V6T 2A3
        Canada
        e-mail: wangsc@triumf.ca

        Pixon.java sets up the Pixon algorithm */

import java.lang.*;
import java.util.*;

public class Pixon {
 final long dataVol = 1000000000;

 Michel michel1, michel2; // two threads
 LMFit  lm;
 Anneal anneal;
 Random rand;

 public static void main(String args[]) {
        Pixon deconv = new Pixon();

        deconv.initialize();
        deconv.Go();
 }      // end of main

 public Pixon () {
        super();
        rand    = new Random();
        michel1 = new Michel(dataVol/2,rand.nextLong());
        michel2 = new Michel(dataVol/2,rand.nextLong());
        lm      = new LMFit(this);
        anneal  = new Anneal(this);
 }      // end of Pixon constructor

 public void Go() {
        for (int i=0; i<3; i++) {
            ImageFit();
            ModelFit();
        }
        ImageFit();
 }

 public void initialize() {
        try {
            michel1.start();
            michel2.start();
            michel1.join();
            michel2.join();
        } catch (InterruptedException e) {}
        merge();
        anneal.initialize(lm.ndata);
 }

 public void merge() {
        for (int i=0; i<lm.ndata; i++) {
            lm.y[i] = michel1.y[i]+michel2.y[i];
            if (lm.y[i] == 0.0) lm.sig[i] = Double.MAX_VALUE;
            else lm.sig[i] = Math.sqrt(lm.y[i]); // Poisson
        }                                         // statistics
        lm.a[1] = (dataVol-michel1.deduction-michel2.deduction)/
                                                   michel1.norm;
 }

 public void ModelFit() {
        anneal.Go(); // find optimal pixel base given parameters
```

```
        System.arraycopy(anneal.model,0,lm.model,0,lm.ndata);
 }

 public void ImageFit() {
        lm.Go(); // find optimal parameters given base
        System.arraycopy(lm.image,0,anneal.image,0,lm.ndata);
 }
}       // end of class
```

Listing 12.1 Pixon.java

`LMFit` (Listing 12.2) is a revision of `SurfaceFit` of Chapter 11. The only modification is that convolution of the objective function as in Eq. (12.20) are applied whenever function values (and its derivatives) are evaluated. Changes in class `Anneal` of Chapter 4 include parameters for the temperature cooling procedure and the way each pixel size is perturbed around the 'center' value in the `perturb()` method.

```
/*      Sun-Chong Wang
        TRIUMF
        4004 Wesbrook Mall
        Vancouver, V6T 2A3
        Canada
        e-mail: wangsc@triumf.ca

        LMFit.java finds best-fit parameters
        given a pixel (bin) base */

import java.lang.*;
import java.util.*;

public class LMFit {

 final int ma = 2; // # of free parameters: the first Michel
                   // parameter and a normalization constant
 double x_0;
 int nx, ny;
 int ndata;
 double[][] x;
 double[] y, sig;
 double[] weight;
 double[] image, model;

 double[] a;
 double[] dyda;
 double[] delta_a, sigmaa;
 double[][] coef;
 double chisqr;
 double flamda;

 double a1, a2, a3;

 Pixon parent;

 public LMFit(Pixon parent) {
        super();

        this.parent = parent;

        x_0     = parent.michel1.x_0; // a constant for Michel

        nx      = parent.michel1.nx;
        ny      = parent.michel2.ny;

        ndata   = nx*ny; // total # of data points
        x       = parent.michel1.x;
```

```
        y       = new double[ndata];
        sig     = new double[ndata];
        weight  = new double[ndata];

        image   = new double[ndata]; // image data
        model   = new double[ndata]; // pixel base
        coef    = new double[2][5];

        a       = new double[ma];  // for free parameters
        sigmaa  = new double[ma];  // for errors
        dyda    = new double[ma];  // for derivatives
        delta_a = new double[ma];

        chisqr  = 0.0;
        flamda  = -1.0; // got to start from -1

        a[0]    = 0.75;
        a1      = 0.0;  // the rest Michel parameters are fixed
        a2      = 1.0;
        a3      = 0.75;

        coef[0][0] = 1./3.;  // beta = 1
        coef[0][1] = 1./3.;
        coef[0][2] = 1./3.;

        coef[1][0] = 0.2;    // beta =2
        coef[1][1] = 0.2;
        coef[1][2] = 0.2;
        coef[1][3] = 0.2;
        coef[1][4] = 0.2;
        // no smearing initially
        for (int i=0; i<ndata; i++) model[i] = 1.0;
}

public void Go() {
        int j=0, imax=6;
        double oldc;

        flamda = -1.0;
        Marquardt();
        oldc = chisqr;
        System.out.println(chisqr+" at iteration = 0");
        do {   // note: you need to interpret outputs based on maximal
            oldc = chisqr;                // posterior probability, not
            Marquardt();                  // on minimal chi-square
            j += 1;
            System.out.println(chisqr+" at iteration = "+j+" "+flamda);
        } while (((oldc-chisqr)/chisqr) > 0.001 && j < imax);

        flamda = 0.0;
        Marquardt();
        for (j=0; j<ma; j++) {
            System.out.println(a[j]+" +- "+sigmaa[j]);
        }
        System.out.println(a1);
        System.out.println(a2);
        System.out.println(a3);
        System.out.println("chi-sq = "+chisqr);
        System.out.println("NDF = "+ndata+" - "+ma);

        for (j=0; j<ndata; j++) image[j] = func(j,a);
}

public double func(int i, double[] a) {
        double tmp,xx,yy;
```

```
        xx = x[i][0];
        yy = x[i][1];
        // theoretic (Michel) distribution
        if (xx <= 1.0 && xx >= x_0) {
           tmp = Math.sqrt(xx*xx-x_0*x_0)*(6.0*xx*(1.0-xx)+
                 4.0/3.0*a[0]*(4.0*xx*xx-3.0*xx-x_0*x_0)+
                 6.0*a1*x_0*(1.0-xx)+
                 a2*yy*Math.sqrt(xx*xx-x_0*x_0)*(2.0*(1.0-xx)+
                 4.0/3.0*a3*(4.0*xx-4.0+Math.sqrt(1.0-x_0*x_0))));
           return a[1]*tmp; // a[1] is a normalization constant
        } else {
           return 0.0;
        }
}

private void dfunc_da(int j, double[] a, double[] da) {
// calculate derivatives numerically. dyda are needed by LM method
        double tmp;

        for (int i=0; i<ma; i++) {
            if (da[i] != 0.0) {
               tmp = a[i];
               a[i] += da[i];
               dyda[i] = convolution(j,a); // func is now convoluted
               a[i] -= (2.0*da[i]);
               dyda[i] -= convolution(j,a);
               dyda[i] /= (2.0*da[i]);
               a[i] = tmp;
            } else
               dyda[i] = 0.0;
        }
}

public double convolution(int i, double[] a) { // Eq. (12.14)
        int j,k;
        double ytmp;

        ytmp = 0.0;

        if (model[i] == 1.0) {
           ytmp = func(i,a);
        } else if (model[i] == 3.0) {
           if (i == 0) {
              for (j=0; j<=1; j++) {
                  ytmp += (func(j,a)*coef[0][j+1]);
              }
           } else if (i == (ndata-1)) {
              for (j=(ndata-2); j<ndata; j++) {
                  ytmp += (func(j,a)*coef[0][j-i+1]);
              }
           } else {
              for (j=-1; j<=1; j++) {
                  ytmp += (func((i+j),a)*coef[0][j+1]);
              }
           }
        } else if (model[i] == 5.0) {
           if (i == 0) {
              for (j=0; j<=2; j++) {
                  ytmp += (func(j,a)*coef[1][j+2]);
              }
           } else if (i == 1) {
              for (j=0; j<=3; j++) {
                  ytmp += (func(j,a)*coef[1][j+1]);
              }
           } else if (i == (ndata-2)) {
              for (j=(ndata-4); j<ndata; j++) {
```

```
                ytmp += (func(j,a)*coef[1][j-i+2]);
            }
        } else if (i == (ndata-1)) {
            for (j=(ndata-3); j<ndata; j++) {
                ytmp += (func(j,a)*coef[1][j-i+2]);
            }
        } else {
            for (j=-2; j<=2; j++) {
                ytmp += (func((i+j),a)*coef[1][j+2]);
            }
        }
    } else System.out.println("Error in local convolution");

    return ytmp;
}

private void Marquardt() {
    int i,j,k;
    double dy, chisq1;
    double[][] alpha = new double[ma][ma];
    double[] beta = new double[ma];
    double[] b = new double[ma];
    Matrix array = new Matrix(ma,ma);
    boolean gradient;

    if (ndata <= ma) {
       chisqr = 0.0;
       System.out.println("Not enough data points");
       return;
    }

    for (j=0; j<ma; j++) {
        beta[j] = 0.0;
        for (k=0; k<=j; k++) alpha[j][k] = 0.0;
    }

    if (flamda < 0.0) {
       flamda = 0.001;
       for (i=0; i<ndata; i++) weight[i] = 1.0/sig[i]/sig[i];
       for (j=0; j<ma; j++) {
           if (a[j] == 0.0) delta_a[j] = 0.01;
           else delta_a[j] = a[j]/100.0;
       }
    }

    chisq1 = 0.0;
    for (i=0; i<ndata; i++) {
        dy = y[i] - convolution(i,a);
        dfunc_da(i,a,delta_a);
        for (j=0; j<ma; j++) {
            beta[j] += (weight[i]*dy*dyda[j]);
            for (k=0; k<=j; k++)
                alpha[j][k] += (dyda[j]*dyda[k]*weight[i]);
        }
        chisq1 += dy*dy*weight[i];
    }
    chisq1 /= (ndata-ma);

    for (j=0; j<ma; j++)
        for (k=(j+1); k<ma; k++) alpha[j][k] = alpha[k][j];

    do {
       for (j=0; j<ma; j++) {
           for (k=0; k<ma; k++) array.M[j][k] = alpha[j][k];
           array.M[j][j] += alpha[j][j]*flamda;
```

```
            }
            try {
                array.inverse();
            } catch (MyMatrixExceptions mme) {
                System.out.println(mme.getMessage());
            }
            if (flamda != 0.0) {
               for (j=0; j<ma; j++) {
                   b[j] = a[j];
                   for (k=0; k<ma; k++) b[j] += (beta[k]*array.M[j][k]);
               }

               chisqr = 0.0;
               for (i=0; i<ndata; i++) {
                   dy = y[i] - convolution(i,b);
                   chisqr += dy*dy*weight[i];
               }
               chisqr /= (ndata-ma);

               if (chisqr > chisq1) {
                  flamda *= 10.0;
                  gradient = true;
               } else {
                  flamda /= 10.0;
                  gradient = false;
               }
            }  else gradient = false;
        } while (gradient == true);  // more gradient

        for (j=0; j<ma; j++) {
            if (flamda == 0.0) {
               sigmaa[j] = Math.sqrt(array.M[j][j]);
            } else {
               delta_a[j] = b[j]-a[j];
               a[j] = b[j];
            }
        }
    }   // end of method
}       // end of class
```

Listing 12.2 LMFit.java

When the spectrum is smeared according to Eq. (12.22), the chi-square value, evaluated on the default pixel base, rises dramatically, signaling a great deviation of the 'experimental' data from the theoretical distribution. This scenario happens in almost every measurement in real world. We are simply simulating it on computers. The best-fit Michel parameters, $\rho, \eta, P_\mu\xi$, and δ are now off their standard (theoretical) values.

When the pixel base is deformed according to the Pixon algorithm during the annealing, the chi-square drops. It's encouraging that with the naive pixel size function of Eq. (12.20), the best-fit Michel parameters, which were shifted by the effect of smearing, are now restored close to the standard values. The distribution of the pixel sizes obtained in the annealing algorithm is shown in Listing 12.3. It is seen that the trend of the increasing pixel size with energy reflects the smearing function of Eq. (12.21) that we have applied. The iterative Pixon deconvolution procedure proved to be powerful.

```
1.0 3.0 1.0 1.0 5.0 1.0 1.0 1.0 1.0 1.0 1.0 1.0 1.0 5.0 1.0 1.0 1.0 1.0
1.0 5.0 1.0 1.0 1.0 1.0 1.0 1.0 1.0 1.0 1.0 1.0 5.0 1.0 1.0 1.0 1.0 5.0
1.0 1.0 1.0 1.0 1.0 1.0 1.0 1.0 1.0 1.0 1.0 1.0 1.0 5.0 5.0 1.0 5.0 1.0
3.0 1.0 1.0 5.0 1.0 5.0 1.0 5.0 5.0 1.0 1.0 1.0 1.0 5.0 1.0 5.0 1.0 1.0
```

```
1.0 1.0 1.0 5.0 5.0 5.0 1.0 1.0 5.0 1.0 1.0 5.0 1.0 1.0 1.0 1.0 5.0 1.0
1.0 1.0 5.0 1.0 1.0 1.0 5.0 1.0 1.0 5.0 1.0 1.0 1.0 1.0 5.0 1.0 5.0 1.0
5.0 1.0 3.0 1.0 1.0 1.0 1.0 5.0 5.0 1.0 1.0 1.0 1.0 1.0 1.0 1.0 1.0 1.0
1.0 1.0 5.0 1.0 1.0 5.0 1.0 1.0 1.0 1.0 5.0 5.0 5.0 1.0 5.0 1.0 1.0 1.0
5.0 1.0 1.0 1.0 5.0 5.0 1.0 5.0 1.0 1.0 1.0 5.0 1.0 1.0 1.0 1.0 5.0 1.0
5.0 1.0 5.0 5.0 1.0 5.0 5.0 5.0 5.0 1.0 1.0 5.0 1.0 5.0 5.0 1.0 5.0 1.0
5.0 5.0 5.0 1.0 5.0 5.0 5.0 5.0 5.0 1.0 5.0 1.0 5.0 5.0 1.0 5.0 5.0 1.0
5.0 1.0 5.0 1.0 5.0 5.0 5.0 5.0 5.0 5.0 5.0 5.0 5.0 5.0 5.0 5.0 1.0 1.0
5.0 5.0 5.0 5.0 1.0 5.0 1.0 5.0 5.0 5.0 5.0 5.0 5.0 5.0 5.0 5.0 5.0 5.0
5.0 5.0 5.0 5.0 1.0 5.0 5.0 5.0 5.0 5.0 1.0 5.0 5.0 5.0 5.0 5.0 5.0 5.0
5.0 5.0 5.0 5.0 5.0 5.0 5.0 5.0 5.0 5.0 5.0 5.0 5.0 5.0 5.0 5.0 5.0 5.0
5.0 5.0 5.0 5.0 5.0 5.0 5.0 5.0 5.0 5.0 5.0 5.0 5.0 5.0 5.0 5.0 5.0 5.0
5.0 5.0 5.0 5.0 5.0 5.0 5.0 5.0 5.0 5.0 5.0 5.0 5.0 5.0 5.0 5.0 5.0 5.0
5.0 5.0 5.0 5.0 5.0 5.0 5.0 5.0 5.0 5.0 5.0 5.0 5.0 5.0 5.0 5.0 5.0 5.0
5.0 5.0 5.0 5.0 5.0 5.0 5.0 5.0 5.0 5.0 5.0 5.0 5.0 5.0 5.0 5.0 5.0 5.0
5.0 5.0 5.0 5.0 5.0 5.0 5.0 5.0 5.0 5.0 5.0 5.0 5.0 5.0 5.0 5.0 5.0 5.0
5.0 5.0 5.0 5.0 5.0 5.0 5.0 5.0 3.0 1.0
```

Listing 12.3 distribution of the output pixel sizes

12.7 Summary

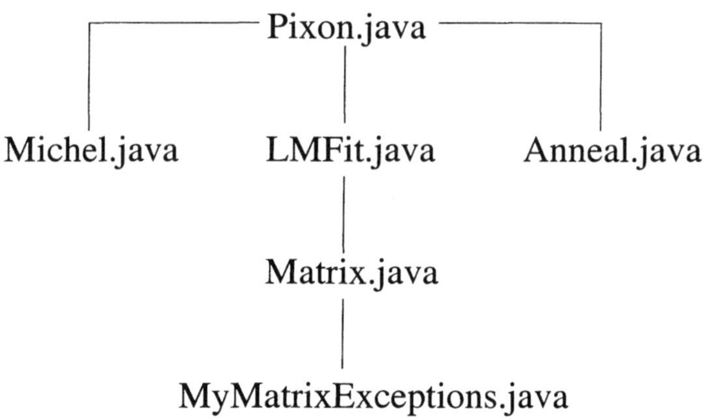

Figure 12.2. Source programs for the Pixon deconvolution algorithm

`Matrix.java` and the associated exceptions class are the same as in Chapter 1. `Anneal.java` of this application comes from that of Chapter 4.

We presented a one dimensional image (spectrum) restoration method based on Bayesian analysis. The source programs involved are shown in Figure 12.2. The posterior probability of the system, which is the product of the likelihood function and the prior probability, was constructed and maximized with respect to two alternate sets of parameters. The first set are the ones (Michel parameters) adjusted in the chi-square minimization by the Levenberg-Marquardt method. The second set are the pixel widths adjusted by the simulated annealing method. The iterative procedure continues until the posterior probability converges.

The spectral data were sampled from a theoretical distribution by Monte Carlo. They were smeared before binned into the histogram. The smearing correlates counts in adjacent bins (pixels). The number of independent bins of the histogram is thus less than the number of 'physical' bins. This insight justifies the Pixon algorithm where a new set of fundamental pixel sizes are sought in order to maximize the posterior probability.

We introduced the principle of maximum entropy to assign prior probabilities with maximum ignorance (therefore, minimum prejudice). Bayesian inference has been exploited as a form of artificial intelligence and found in many latest personalized software.

12.8 References and Further Reading

The Pixon algorithm was introduced in, R.K. Pina and R.C. Puetter, "Bayesian Image Reconstruction: The Pixon and Optimal Image Modeling", *Publication of the Astronomical Society of the Pacific*, **105** (1993) 630-637

The entropy was defined in, for the study of efficient communications, C.E. Shannon, "A Mathematical Theory of Communication", *Bell System Technical Journal*, **27** (1948) 379-423

An introduction to Bayesian data analysis can be found in, G.L. Bretthorst, "Bayesian Spectrum Analysis and Parameter Estimation" in *Lecture Notes in Statistics* **48**, Springer-Verlag, New York (1988)

An online clearinghouse for the Bayesian approach to statistical inference is, `http://astrosun.tn.cornell.edu/staff/loredo/bayes/`, where enormous Bayes resources, such as naive Bayesian learning and belief network, can be reached.

The following web site is dedicated to applications of entropy in such fields as information and coding theory, dynamical systems, logic and the theory of algorithms, statistical inference, and biology: `http://www.informatik.uni-trier.de/~damm/Lehre/InfoCode/entropy.html`

Chapter 13

GRAPHICAL MODEL

Complexity in nature or biology results more from the structure of the system than from some 'magic' parameter values in the system. Examples are transcriptional networks of genes and the Internet, both of which are resilient to random attacks. Network structures have been studied by graphical models.

Many real-time applications are critical on the speed of data processing. Furthermore, the data in some applications are arriving sequentially. Therefore, an algorithm which fits the data progressively and performs as well as the chi-square minimization method is highly desirable. Both requirements are met by the technique of Kalman filter, which is a special case of graphical models. Prerequisites for Kalman filtering can be greatly relaxed when we use the method of H infinity filter.

13.1 Directed Graphs

A graph (or network) contains nodes and arcs (between the nodes). Figure 13.1 (a) shows an example of a directed graph. A node is associated with a random variable and an arc between two nodes establishes a dependence relationship between them. For instance, circles (nodes) in Figure 13.1 (a) can represent genes and the associated random variables represent levels of gene expression (activity). Figure 13.1 (a) then reads: gene 1 regulate gene 2, gene 2 regulate gene 3, ..., etc. Modeling a dependence relationship can be achieved via a conditional probability density. Thus, the arc pointing from gene 1, g_1, to gene 2, g_2, carries a finite $P(g_2|g_1)$. Once we have the structure (arcs) of a graph, we can write down the joint probability density of the structure as a product of the conditional probability densities. *Learning* in graphical models refers to finding the structure (and the embedded parameter values) whose joint probability density is maximal.

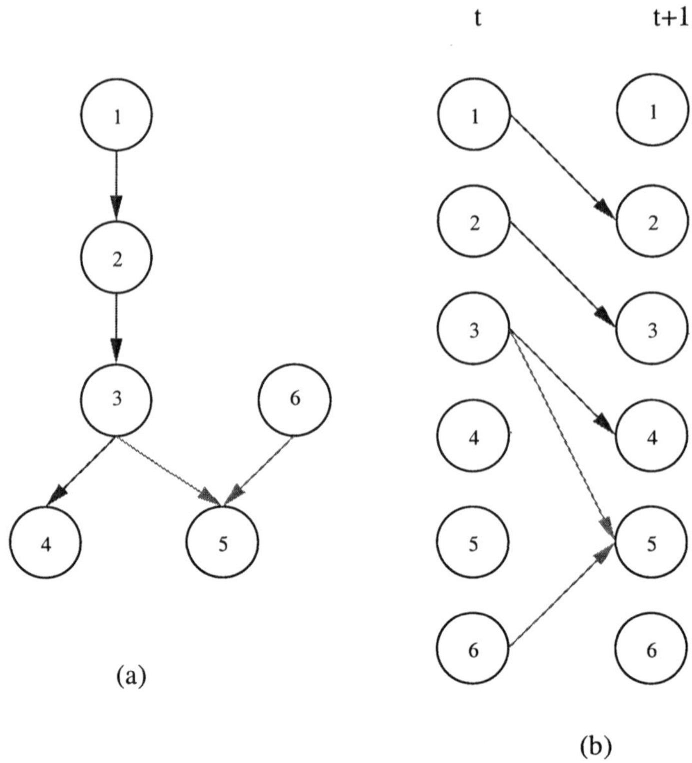

Figure 13.1. An example of directed graph. (b) is the same as (a) except additional time information.

When learning graphs from time series data, we simply repeat the *static* graphs and draw arcs between nodes across time. Figure 13.1 (b) is the dynamical version of 13.1 (a). In this case, an arc comes with a finite transitional probability density such as $P(g_i(t+1)|g_j(t))$.

13.2 Bayesian Information Criterion

To give a concrete example, assume the interdependence of stock indexes is modeled by a directed graph and that changes in stock index x_i are affected by other stock indexes x_j in the following way,[1]

$$\frac{dx_i}{dt} = \alpha_i \prod_j x_j{}^{w_{ji}}, \tag{13.1}$$

[1] x could have been the logarithm of a stock index. For a detailed modeling of financial assets, please refer to Sections 7.6 and 7.7.

where α_i is a positive constant and the product is over the set of indexes that affect index i. If $w_{ji} > 0$ ($w_{ji} < 0$), then an increase in index j drives index i up (down). The discrete-time form of Eq. (13.1) is,

$$x_i(t+1) = \alpha_i \prod_j x_j(t)^{w_{ji}} + x_i(t) + \epsilon_i(t), \tag{13.2}$$

where we have included $\epsilon_i(t)$ which is the volatility of index i at time t. For simplicity, let's assume that $\epsilon_i(t)$'s are independent of time and that they are Gaussian distributed with mean zero and standard deviation, σ_i: $\epsilon_i(t) = \epsilon_i = G(0, \sigma_i)$. The task is to find the set of arcs and the associated w's and α's from time series observations of the stock indexes.

Once we have a candidate structure of the stock index regulation network, the conditional probability density, $p(x_i, t; \theta_i)$, of any stock index i in the network at time t can be readily written down as,[2]

$$\begin{aligned} p(x_i, t; \theta_i) &\equiv p(x_i, t; \alpha_i, w_{ji}) \\ &= G\Big(x_i(t) - \alpha_i \prod_j x_j(t-1)^{w_{ji}} - x_i(t-1), \sigma_i\Big) \\ &= \epsilon_i(t) = \epsilon_i. \end{aligned} \tag{13.3}$$

The probability density, $P(S|\mathbf{x}; \theta)$, of a candidate network structure, S, given the observational data and parameters becomes

$$P(S|\mathbf{x}; \theta) = \prod_{t=1}^{T-1} \prod_{i=1}^{n} p(x_i, t; \theta_i), \tag{13.4}$$

where there are n indexes in the network and T observations at time points 0, 1, $\cdots$, and $T-1$. Since we don't know precise values of α's, w's, and σ's *a priori*, to take them into account unbiasedly, we integrate $P(S|\mathbf{x}; \theta)$ over possible ranges of the parameters. This marginalization can prove to be intractable and we employ the approximation that approaches the logarithm of $P(S|\mathbf{x})$ as long as $T - 1$ is large enough,

$$\text{score}(S) = \log\big(P(\mathbf{x}|\hat{\theta}, S)\big) - \frac{d}{2}\log(T-1), \tag{13.5}$$

where $\hat{\theta}$ are the parameters that maximize the likelihood function $P(\mathbf{x}|\theta, S)$ and the second term, with d equal to the number of parameters in the network structure, is a penalty against structure complexity. σ's can be estimated from

[2]Prior probability distributions on the α_i, w_{ji}, and σ_i are assumed uniform. If we have prior knowledge about their distributions, we can include them here.

data. The first term of the score metric is, apart from a normalization constant, the minimum of the chi-square function. The task of stock regulation network reconstruction is cast into search for the structure, together with the embedded parameters, that scores highest. The importance of the form of the score function, called Bayesian information criterion, is that we seek parsimonious structures that fit the data. Other score functions of the same virtue, such as Akaike information criterion and its finite size corrections, are available. They differ in the weight on the penalty relative to the chi-square term.

We can use genetic algorithm of Chapter 6 to search for the structure. Embedded in the genetic algorithm can be the simulated annealing of Chapter 4 that finds optimal values of the parameters in the arcs.

13.3 Kalman Filter

Consider a stream of signals emitted from a couple of satellites in space. A global positioning system (GPS) on the ground receives the signals from which the coordinates of the GPS receiver's location are updated. Similarly, a computer robot tracks a moving object (for example, a vehicle) via an on-board camera. A canonical problem of the like is the tracking of guided missiles.

To proceed, we write down a *system equation* relating the state x of the system at time t_k to its state at a previous time t_{k-1},

$$x_k = f_{k-1}(x_{k-1}) + w_{k-1}, \tag{13.6}$$

where subscripts index time. $f_{k-1}(x_{k-1})$ is the propagator which transforms the state of the system at time t_{k-1} to the next instance of time t_k. w_{k-1} is called *process noise* or *process disturbance.* and is intrinsic to the dynamics of the system. To describe the state of, say, a vehicle, x can include the (3-dimensional spatial) coordinates and the velocity. $f(x)$, a matrix, then models how the state vector changes under acceleration, and so on.

The sensing device hardly measures the state directly. The video stream of the camera, for instance, outputs no more than 2-dimensional images of the scene, which reads no quantities such as velocity. For the sake of simplicity, a vehicle is represented by a 2-dimensional bounding box in the image. We then write down the *measurement equation*,

$$m_k = h_k(x_k) + \epsilon_k, \tag{13.7}$$

where m_k are the measured data (signals) at time t_k and $h_k(x_k)$ transforms the state vector of the system, x_k, into the measurement vector (from a 3-dimensional bounding box into a 2-dimensional one, for example). ϵ_k are the associated measurement error.

It is now clear that a Kalman filter is a graphical model whose structure is known and fixed (Figure 13.2). The task remains to estimate the parameters

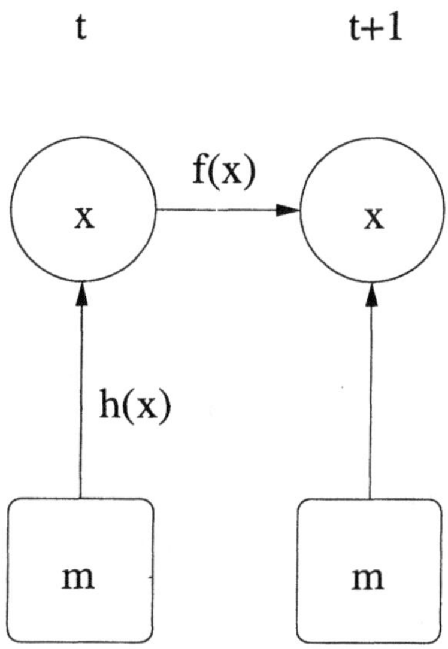

Figure 13.2. A Kalman filter

in the model $f(x)$ which holds vital information such as where the missile is heading for and how fast it is. The nice feature of a Kalman filter is that parameter estimation is accomplished in real-time, which justifies its indispensability in time critical applications.

13.4 A Progressive Procedure

The filter has its origin in probability. What is sought is the state x_{k+1} whose probability is maximum given all previous measurements, $m_k, m_{k-1}, \cdots, m_1$,

$$P(x_{k+1}|m_k, m_{k-1}, \cdots, m_1) \equiv P(x_{k+1}|M_k). \tag{13.8}$$

By Bayes theorem, $P(A|B)P(B) = P(B|A)P(A)$, and the property of a Markov process,

$$P(x_k|x_{k-1}, x_{k-2}, \cdots, x_0) = P(x_k|x_{k-1}), \tag{13.9}$$

it can be shown that,

$$P(x_{k+1}|M_k) \propto \int dx_k\, P(x_k|M_{k-1})\, P(m_k|x_k)\, P(x_{k+1}|x_k). \tag{13.10}$$

Note the recursive nature of the formula. It tells the important fact that the state of the system can be efficiently updated from a previous estimate whenever new data are available, without recomputing everything. It is the progressive nature of the algorithm that makes Kalman filters adequate in most time-critical applications.

Equation (13.10) can be expressed in an operative way,

$$P(x_{k+1}|M_k) = \int dx_k \, P(x_{k+1}|x_k) \, P(x_k|M_k) \tag{13.11}$$

and

$$P(x_k|M_k) \propto P(m_k|x_k) \, P(x_k|M_{k-1}). \tag{13.12}$$

Equation (13.11) predicts state system before measurement while Eq. (13.12) updates state given measurement data. Maximizing Eq. (13.11), (13.12) gives, therefore, the *a priori*, *a posteriori* estimate of the state, respectively.

To write down the Kalman updating formula for the state vector, more about the error has to be specified. If the process noise in Eq. (13.6) and the measurement error in Eq. (13.7) are Gaussian distributed with zero means and independent of each other, x_k's and m_k's are all Gaussian random variables. Closed form and handy formulas exist, making the manipulation and maximization of Eq. (13.10) tractable. For example, the likelihood, $\mathcal{L}$, of Eq. (13.11) can be defined in the following form,

$$\mathcal{L} = \exp\left\{-\frac{1}{2}\left[\frac{d_m - d_f(x+\Delta x)}{\sigma}\right]^2\right\} \exp\left(-\frac{1}{2}\Delta x^T Q^{-1} \Delta x\right), \tag{13.13}$$

where Δx is the improvement to the parameters we are seeking. d_m are the measurement with error σ. d_f transforms the parameters to measurement. Q is the covariance matrix of the parameters. The first exponential weighs the measurement data while the second constrains the range of possible Δx. The goal is to find Δx which maximizes the likelihood. The resulting filtering (updating) formula is,

$$x_k = x_k^{k-1} + K_k(m_k - H_k x_k^{k-1}), \tag{13.14}$$

where x_k^{k-1} stands for the estimate of x at time t_k using all measurements up to time t_{k-1}. x_k^{k-1} and x_k are the *a priori* and *a posteriori* estimate of the state, respectively. The quantity in the parentheses is the difference between the measurement and the *a priori*, and is called *residual*. K_k is usually called gain matrix and is defined as,

$$K_k = \frac{C_k^{k-1} H_k^T}{V_k + H_k C_k^{k-1} H_k^T}, \tag{13.15}$$

where superscript ^{T} means transpose of the matrix. C_k^{k-1} is the *a priori* estimate error covariance matrix,

$$C_k^{k-1} = \overline{e_k^{k-1} e_k^{k-1T}} = \overline{(x_k^{k-1} - \hat{x}_k)(x_k^{k-1} - \hat{x}_k)^T}, \tag{13.16}$$

where $\hat{x}_k$ is the true value of the state. The overline stands for average. V_k in the gain matrix formula is simply the covariance matrix of the measurement error,

$$V_k = \overline{\epsilon_k \epsilon_k^T}. \tag{13.17}$$

A pedagogical interpretation of Eq. (13.14) is in order. Observe that,

$$K_k \approx \begin{cases} \frac{1}{H_k}, & \text{as} \quad V_k \to 0 \\ 0, & \text{as} \quad C_k^{k-1} \to 0 \end{cases} \tag{13.18}$$

which says that the algorithm weighs more on measurement when the measurement error is small. It, on the other hand, takes less account of measurement while the *a priori* estimate error is small. In other words, a tug of war takes place between measurement error and process noise, and a balance is prescribed by the gain matrix: Eq. (13.15).

The state vector evolves according to the system equation Eq. (13.6). The corresponding propagation for the *a priori* estimate error covariance matrix is,

$$C_k^{k-1} = F_{k-1} C_{k-1} F_{k-1}^T + Q_{k-1}, \tag{13.19}$$

where Q_{k-1} is the process noise covariance matrix,

$$Q_{k-1} = \overline{w_{k-1} w_{k-1}^T}. \tag{13.20}$$

Finally, the *a posteriori* estimate error covariance is calculated by,

$$C_k = (1 - K_k H_k) C_k^{k-1}. \tag{13.21}$$

13.5 Kalman Smoother

Imagine that a series of N measurements is finished. With the Kalman formulas, the state vector at the final time t_N used all the measurement data up to m_N. It therefore gives the best-fit estimate of the parameters. However, in some off-line situations one is interested in the parameter estimate at an earlier time t_k, $k < N$. The previous filtering result of x_k is not the optimal one since its calculation does not include any measurement data after m_k. To make use of all the information, the same Kalman filter can be applied again, with the filter running backwards. The best estimate is then the (weighted) average of the two: one forward, the other backward.

A set of formulas called Kalman *smoother* automatically takes into account the averaging. They are ($n > k$),

$$x_k^n = x_k + A_k (x_{k+1}^n - x_{k+1}^k), \tag{13.22}$$

for the smoothed state vector, where

$$A_k = C_k F_k^T (C_{k+1}^k)^{-1} \tag{13.23}$$

is the smoother gain matrix. The covariance matrix of the smoothed state vector is,

$$C_k^n = C_k + A_k(C_{k+1}^n - C_{k+1}^k)A_k^T. \tag{13.24}$$

Note the matrix inversion in the above equation. The size of the covariance matrix is 5 by 5 if the size of the state vector is 5.

13.6 Initialization of the Filter

Since the procedure is recursive, we guess an initial state for the algorithm to take off. The usual practice is to use the origin of the state vector as the starting state. For example, the origin of the coordinate system and a nominal velocity in the vehicle tracking application. In a more sophisticated application, there is a specialized 'first guess' routine whose output feeds to the Kalman filter. For example, a pattern recognition program scans the video data and identifies the bounding box representing the vehicle.

The other important job is to initialize the *a priori* estimate error covariance matrix, C_0. The usual choice is to set it as large as possible if you know nothing of the initial state. Then, according to Eq. (13.18), the initial guess respects more of the measurement and yields to the measurement. The rationale is that Kalman filters are Bayesian. If you are not confident of your first guess (prior), you put less weights on your initial error covariance matrix by using larger values. On the other hand, if you are sure, then you use smaller values in the error covariance matrix. In the former case, however, you waste the first few measurements since the Kalman is now working on finding the proper covariance matrices instead of doing estimations of the state vector. (If the number of measurements is more than enough, this is not a concern.) In the latter case, too small a covariance matrix might bias the Kalman. Namely, the filter is too stubborn to make any change in the initial guess even given the evidence (the measurement). Therefore, any knowledge of the system under consideration is always helpful.

13.7 Helix Tracking

After introducing the ideas and general formulas for the Kalman method, we demonstrate its application to charged particle tracking. In experimental high-energy and astro particle physics, charged particles, prepared in the cyclotron in a laboratory or bombarding Earth from outer space, are tracked by *wire chambers* in a static magnetic field. A charged particle, entering the region of a uniform magnetic field, precesses around the direction of magnetic field. Its velocity component along the field is, however, not changed. The trajectory is therefore a *helix*. Figure 13.3 shows such a helical path. Recall that Earth is a gigantic magnetic dipole. The field lines of the dipole help keep most low (to intermediate) energy cosmic rays (charged muons) from reaching the ground.

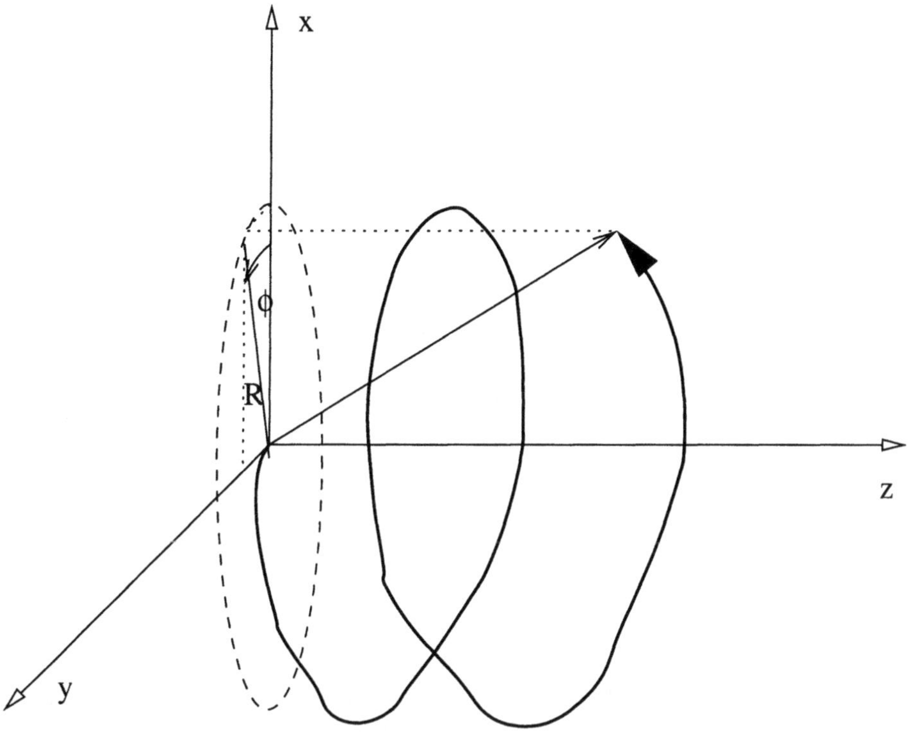

Figure 13.3. Track of a charged particle in a uniform magnetic field in the z direction. The direction of winding depends on the sign of the charge

Imagine that, in the region of the particle tracks, there are planes of parallel wires. In one plane, wires are strung vertically (say, x direction). In the next adjacent plane, wires go horizontally (y direction). The x planes interleave the y planes, as shown in Figure 13.4. When a charged particle comes close to the wire, the wire fires. When the experimenter records the fired wires resulting from particle crossing, she can reconstruct the tracks from the coordinates of the fired wires.

Return to the Kalman tracking. The state vector of the system of our choice is,

$$P = (x, y, R^{-1}, t, \phi), \tag{13.25}$$

where x, y are the coordinates of the track (z is known, which is the location of the wire planes). R is the radius of the circle obtained when the helix is projected on the $x - y$ plane (see Figure 13.3). t is the ratio of the longitudinal to transverse momentum of the particle. ϕ is the azimuthal angle, measured from the x axis, as the particle advances. From Figure 13.3, it can be seen that

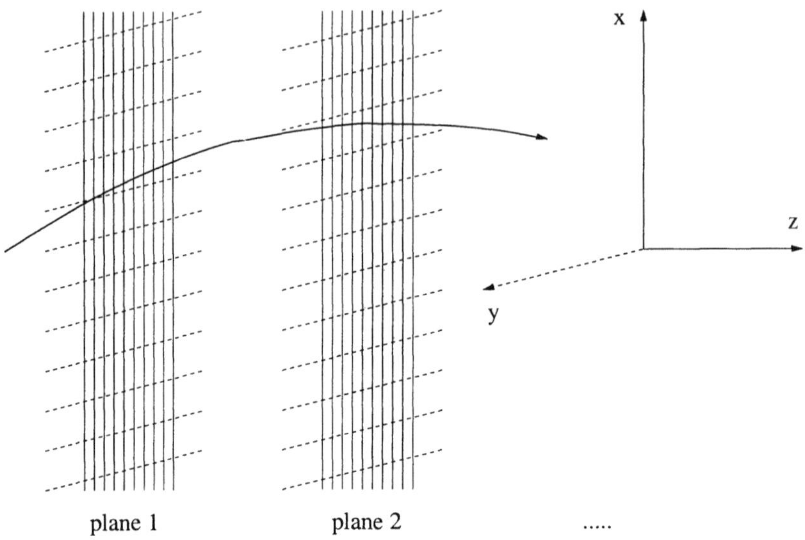

Figure 13.4. A wire chamber. A plane consists of two closely stagged sets of parallel wires going in different directions. From the fired wires which define the x and y and the plane numbers, we get the spatial coordinates of the particle trajectory.

the state vector at the next (downstream) wire plane, P', is related to that of the current plane by,

$$\begin{aligned} x' &= x + R' \cos\phi' - R\cos\phi \\ y' &= y + R' \sin\phi' - R\sin\phi \\ \phi' &\approx \phi + \frac{z' - z}{t} \cdot \frac{2}{R' + R} \\ R'^{-1} &\approx R^{-1} - R^{-1} \cdot \frac{\Delta E}{p} \\ t' &= t \end{aligned} \tag{13.26}$$

where primed (unprimed) quantities refer to those of the next (current) plane. ΔE is the energy loss of the particle after traveling a distance between two planes. If, for simplicity, $\Delta E = 0$, then the above formulas become exact. The 5×5 transport matrix f_k is therefore obtained by,

$$f_{kij} = \left.\frac{\partial P'_i}{\partial P_j}\right|_k , \tag{13.27}$$

where P_i is the ith component of the state vector. Note that subscript k in the formulas indexes wire planes instead of time in the present case. The process noise, w_k, in the system equation of Eq. (13.6) now comes from physics processes such as multiple scattering and energy loss of the particle traversing the space filled with supporting materials of the wire chambers, ionization gas, cathode foils, and the like. To be complete, the covariance matrix Q of Eq. (13.20) due to multiple scattering is,

$$Q = \begin{pmatrix} R^2t^2+y^2 & -xy & -t^2\cos\phi & -xt(1+t^2) & -y(1+t^2) \\ -xy & R^2t^2+x^2 & -t^2\sin\phi & -yt(1+t^2) & x(1+t^2) \\ -t^2\cos\phi & -t^2\sin\phi & t^2(R^{-1})^2 & R^{-1}t(1+t^2) & 0 \\ -xt(1+t^2) & -yt(1+t^2) & R^{-1}t(1+t^2) & (1+t^2)^2 & 0 \\ -y(1+t^2) & x(1+t^2) & 0 & 0 & 1+t^2 \end{pmatrix} \theta_o^2, \tag{13.28}$$

where θ_o is some constant depending on the material of the detector. These noise terms can be calculated and included on the fly from the current plane to the next predicted plane in the fitting. This is the very reason why particle physicists favor the Kalman method over traditional chi-square fits.

The h_k matrix in the measurement equation, Eq. (13.7), now reads,

$$h_k = \begin{pmatrix} 1 & 0 & 0 & 0 & 0 \\ 0 & 0 & 0 & 0 & 0 \end{pmatrix}, \tag{13.29}$$

or

$$h_k = \begin{pmatrix} 0 & 0 & 0 & 0 & 0 \\ 0 & 1 & 0 & 0 & 0 \end{pmatrix}, \tag{13.30}$$

depending on whether the kth plane measures the x or y coordinate of the track. The measurement error, ϵ_k, specifies how precise the wire chamber detector measures position.[3]

13.8 Buffered I/O

When the paces of different computer tasks do not match, it is advantageous to spawn a thread for the slow job so that the program does not waste time waiting for the result of the slow job. We demonstrate such a technique in *event* reading and subsequent Kalman fitting of the event.

[3]The measurement error of a wire chamber detector is the spacing between adjacent parallel wires if no drift time information is used.

First of all, class Event (Listing 13.1) defines the set of data which constitutes a track event. The data basically hold information about which wires of which plane fired, and so on. Note the implementation of the Cloneable interface and clone() method in Event.java. It ensures a *deep* copy of an event data. A buffer of 10-event size is then created in class Buffer (Listing 13.2).

```
/*      Sun-Chong Wang
        TRIUMF
        4004 Wesbrook Mall
        Vancouver, V6T 2A3
        Canada
        e-mail: wangsc@triumf.ca

        Event.java stores information of a track event */

class Event implements Cloneable {
 int Num_Planes;       // total no of planes in the detector system
 int Max_Num_Hits;     // max no of hits
 int[] n_Hits_Plane;   // no of hits on the plane
 int[] wires;          // no of fired wires on the plane
 int[][] TDC_r;        // rising time of the hit
 int[][] TDC_t;        // trailing time of the hit
 int[][] Channel;      // which wires (TDC channel) are hit

 public Event() {
        Num_Planes = 8;           // change while upgrading
        Max_Num_Hits = 80;        // capacity of TDC 1877
        n_Hits_Plane = new int[Num_Planes];
        wires    = new int[Num_Planes];
        Channel  = new int[Num_Planes][Max_Num_Hits];
        TDC_r    = new int[Num_Planes][Max_Num_Hits];
        TDC_t    = new int[Num_Planes][Max_Num_Hits];
 }

 public Object clone() {
        Event copy = (Event) super.clone();
        copy.n_Hits_Plane = new int[Num_Planes];
        copy.wires    = new int[Num_Planes];
        copy.Channel  = new int[Num_Planes][Max_Num_Hits];
        copy.TDC_r    = new int[Num_Planes][Max_Num_Hits];
        copy.TDC_t    = new int[Num_Planes][Max_Num_Hits];

        // deep copying
        for (int j=0; j<Num_Planes; j++) {
            copy.n_Hits_Plane[j] = n_Hits_Plane[j];
            copy.wires[j] = wires[j];
            for (int i=0; i<Max_Num_Hits; i++) {
                copy.Channel[j][i] = Channel[j][i];
                copy.TDC_r[j][i] = TDC_r[j][i];
                copy.TDC_t[j][i] = TDC_t[j][i];
            }
        }
        return copy;
 }
}       // end of class
```

Listing 13.1 Event.java

```
/*      Sun-Chong Wang
        TRIUMF
```

```
        4004 Wesbrook Mall
        Vancouver, V6T 2A3
        Canada
        e-mail: wangsc@triumf.ca

        Buffer.java hosts events to be processed. Two threads,
        Reader and Loader, access this class in the following scheme,
        raw_data -> Reader -> Buffer -> Loader -> pre-processor */

import java.io.*;

class Buffer {
 final int QUEUE_SIZE = 10;     // 10 event in size
 final int LAST = QUEUE_SIZE - 1;
 Event the_queue[] = new Event[QUEUE_SIZE];
 int the_no_in_queue = 0;
 int the_head = 0;
 int the_tail = 0;

 Tracker parent;

 public Buffer(Tracker parent) {
        this.parent = parent;
 }

 public synchronized Event get() {
        while (the_no_in_queue <= 0) {
               try {
                   wait();
               } catch (InterruptedException e) {}
        }
        Event res = the_queue[the_head];
        the_head = the_head == LAST ? 0 : the_head+1;
        the_no_in_queue--;
        notifyAll();
        return res;
 }

 public synchronized void put(final Event value) {
        while (the_no_in_queue >= QUEUE_SIZE) {
               try {
                   wait();
               } catch (InterruptedException e) {}
        }
        if (value != null) {
        the_queue[the_tail] = (Event) value.clone();
        } else { the_queue[the_tail] = value; }
        the_tail = the_tail == LAST ? 0 : the_tail+1;
        the_no_in_queue++;
        notifyAll();
 }
}       // end of class Buffer
```

Listing 13.2 Buffer.java

Data are read from files or hardware (the data acquisition system) by class `Reader` (Listing 13.3) which `extends Thread`. The data are then put event by event to the buffer by `Buffer`'s `put()` method until the buffer is full. If the reading is idled (due to networking, for example) or the buffer is full, the program can work on Kalman fitting or other system tasks. Class `Loader` (Listing 13.4) which also `extends Thread`, on the other hand, gets an event from the buffer by `Buffer`'s `get()` method and feeds it to the Kalman tracking until the buffer is empty. Note the keyword `synchronized` in front of the buffer methods `get()` and `put()`. It modifies the property of the class so that whenever the synchronized method is called by methods of other objects, the buffer is locked up and no further methods can have access to the buffer. This prevents the case where an event is read (by a thread) while writing to the buffer is still underway (by the other thread). The event read could be incomplete without the lock. Care has to be taken when programming multi-threads.

```
/*      Sun-Chong Wang
        TRIUMF
        4004 Wesbrook Mall
        Vancouver, V6T 2A3
        Canada
        e-mail: wangsc@triumf.ca

        Reader.java reads and unpacks events from the
        data acquisition system. Since this is system
        dependent, the purpose of this file is to show
        buffered I/O. The user needs to fill up the
        void in readEvent() to suit her application */

import java.io.*;
import java.lang.*;

class Reader extends Thread {
 final int Num_Planes = 8;
 Buffer the_buffer;
 Event  the_event;
 final int EOF = -1;
 int N = 0;
 private Runtime r = Runtime.getRuntime();

 public Reader(Buffer buffer) {
        super("Reader");
        the_event  = new Event();
        the_buffer = buffer;
 }

 public void run() {
        try {
             FileInputStream fis = new
                                FileInputStream(FileDescriptor.in);
             InputStreamReader br= new InputStreamReader(fis);
             BufferedReader re   = new BufferedReader(br);
             StreamTokenizer sto = new StreamTokenizer(re);

             sto.eolIsSignificant(false);

             N = 0;
             sto.nextToken();
             while (sto.ttype != StreamTokenizer.TT_EOF) {
                   try {
                       readEvent(sto);
                   } catch (TrackerExceptions te) {
                   } catch (StringIndexOutOfBoundsException siobe) {
                       System.out.println(siobe.getMessage()+
                                                " at N = "+N);
                   }
             }     // end of outer while

             System.out.println("total : " + r.totalMemory());
             System.out.println("free  : " + r.freeMemory());
             fis.close();
        } catch (IOException e) {
             System.out.println("R] I/O Error   : " + e.getMessage());
        }
        the_buffer.put(null);
```

```
 }       // end of method run

 public void readEvent(StreamTokenizer sto) throws TrackerExceptions {
        int i,j=0;

        try {
          while (nextword(sto).equals("FBU1")) {  // fast bus 1
               N += 1;
               j = 0;
               while (j < Num_Planes) {  // loop over planes
/*                     // user fill up this part
                       // this part is highly system dependent.
                       // you need to fill the following entries,
                       the_event.n_Hits_Plane[j] =
                       the_event.Channel[j][i] =
                       the_event.TDC_t[j][i]    =
                       the_event.TDC_r[j][i]    =
                       the_event.wires[j] =
                       // and exit this event reading, if something
                       // abnormal happened, by
                       throw new TrackerExceptions("abnormal TDC edge");
                       j += 1;
*/
               }       // plane while loop
               the_buffer.put(the_event);
               }       // an event built
          }            // end of inner while
        } catch (StringIndexOutOfBoundsException siobe) {
        }              // try-catch
 }
}       // end of class Reader
```

Listing 13.3 Reader.java

```
/*      Sun-Chong Wang
        TRIUMF
        4004 Wesbrook Mall
        Vancouver, V6T 2A3
        Canada
        e-mail: wangsc@triumf.ca

        Loader.java gets events from the buffer and sends
        the event to the DataBase class for pre-processing.
        If the data is clean, the event is reconstructed by
        the Kalman method. If successful, the reconstructed
        track is plotted */

import java.io.*;

class Loader extends Thread {
 Buffer the_buffer;
 Tracker parent;
 Event the_event;
 final int EOF = -1;
 int n;

 public Loader(Tracker parent) {
        super("Loader");
        this.parent = parent;
        the_buffer = parent.buffer;
 }

 public void run() {
        the_event = the_buffer.get();
        while (the_event != null) {
```

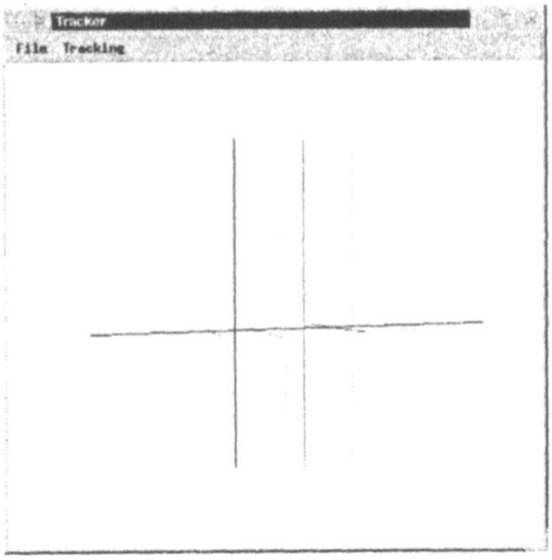

Figure 13.5. Through going particles tracked by Kalman method. Colored lines are fired wires while the black line is the result of Kalman tracking.

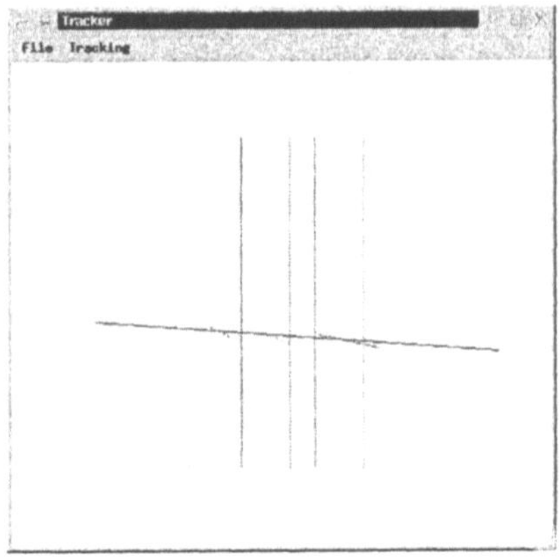

Figure 13.6. Through going particles tracked by Kalman method

```
               parent.db.booking(the_event);  // pre-processing
               if (parent.db.goTracking == true) {
                  parent.kfdlg.tracking.Go();
                  if (parent.kfdlg.update() == true) {
                     parent.rendering.go(the_event);
                     parent.rendering.eventToDraw(the_event);
                     parent.rendering.notifyObservers(
                                    parent.plotting3D);
                  } // plotted after successfully reconstructed
               }    // event data is clean and tracked
               the_event = the_buffer.get();
           }    // end of while
   }       // end of method run
}          // end of class Loader
```

Listing 13.4 Loader.java

13.9 The Kalman Code

Class Loader feeds an event to a database class where it is pre-processed. This procedure throws out bad events, which contain noisy firings. Clean events are then sent to the Kalman fitter before necessary translations. Listing 13.5 shows the Kalman class that tracks straight-lines as well as helices.

```
/*      Sun-Chong Wang
        TRIUMF
        4004 Wesbrook Mall
        Vancouver, V6T 2A3
        Canada
        e-mail: wangsc@triumf.ca

        Kalman.java implements both the Kalman
        filter and smoother */
```

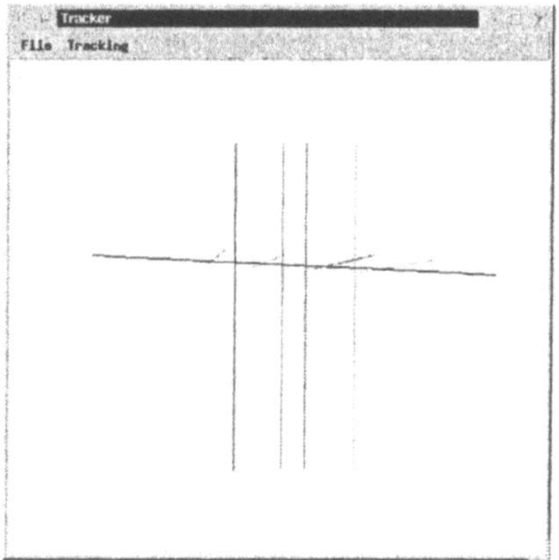

Figure 13.7. Through going particles tracked by Kalman method

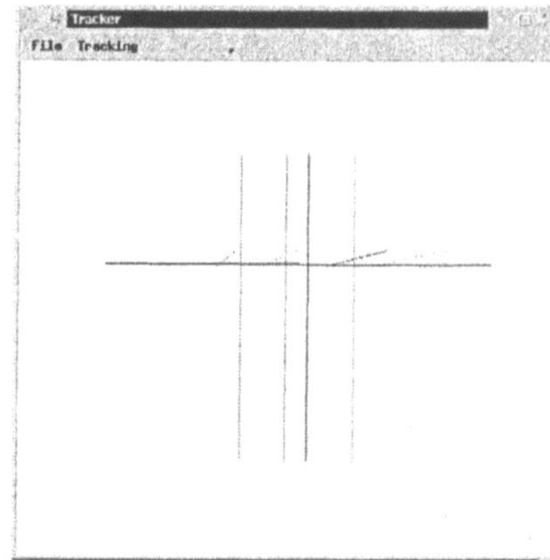

Figure 13.8. Through going particles tracked by Kalman method

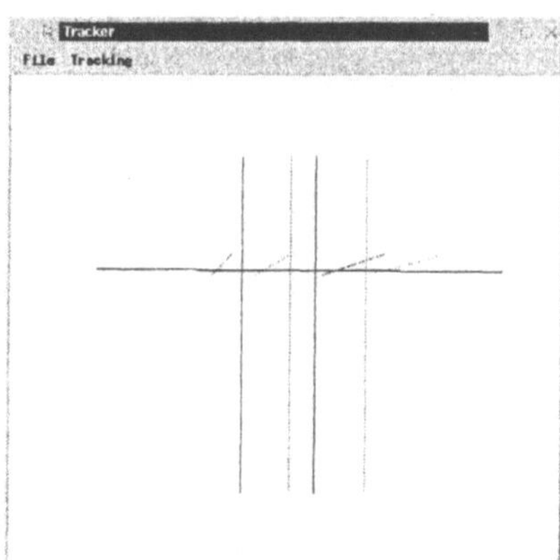

Figure 13.9. Through going particles tracked by Kalman method

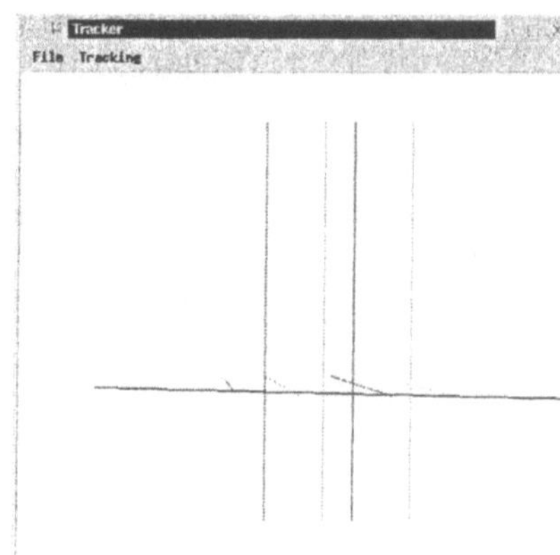

Figure 13.10. Through going particles tracked by Kalman method

```
import java.lang.*;
import java.util.*;

public class Kalman {
 final int  Num_Planes = 8;

 int        iteration, iteration_max;
 int        grandIteration;
 double     Chi2sTotal, epsilon;
 double     x, y, dxdz, dydz;
 double     X2;

 double[]   xw, yw, z;
 double[]   z0;
 int[]      nHits;
```

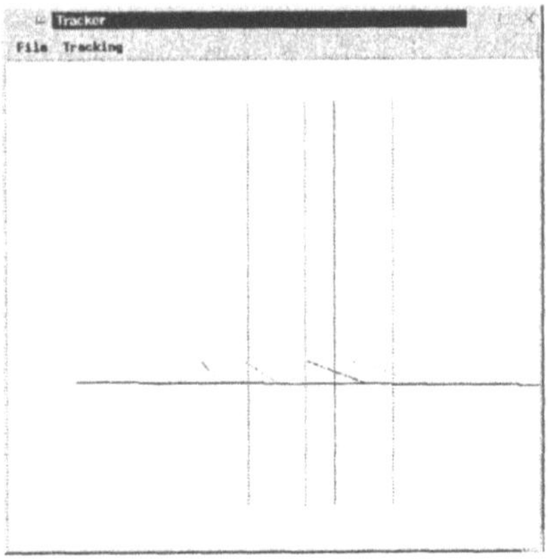

Figure 13.11. Through going particles tracked by Kalman method

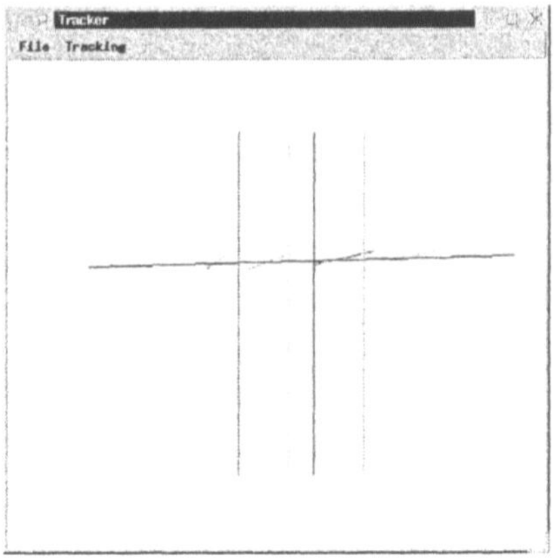

Figure 13.12. Through going particles tracked by Kalman method

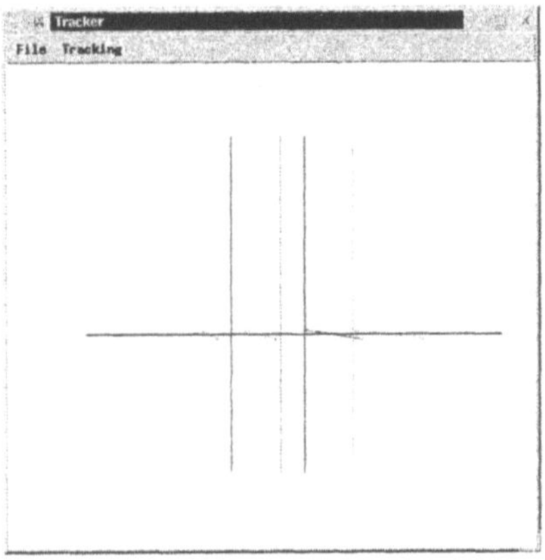

Figure 13.13. Through going particles tracked by Kalman method

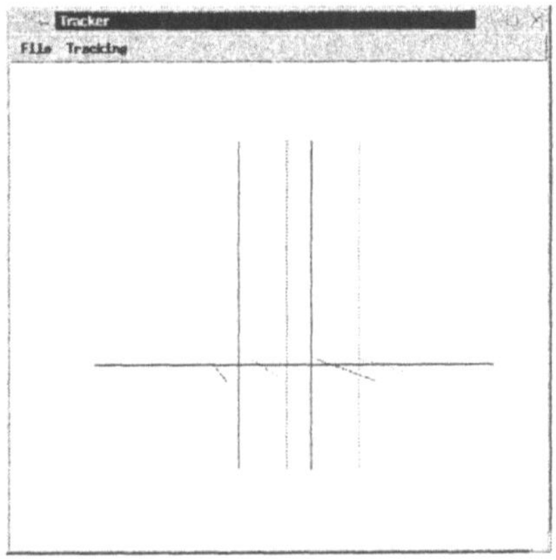

Figure 13.14. Through going particles tracked by Kalman method

```
Matrix V;  // measurement covariance matrix
Matrix As; // smoother gain matrix
Matrix rp, rf, rs; // predicted, filtered, smoothed residuals
Matrix Kf; // filter gain matrix
Matrix Rs; // covariance matrix of smoothed residuals

Matrix G,tmp1,tmp2,tmp3,tmp4;
Matrix reducedCp, inverseTmp, Cp_inverse;

Matrix[] xp    = new Matrix[Num_Planes]; // predicted state vector
Matrix[] Cp    = new Matrix[Num_Planes]; // prediction covariance
Matrix[] xm    = new Matrix[Num_Planes]; // measurement vector
Matrix[] H     = new Matrix[Num_Planes]; // measurement transformation
Matrix[] xf    = new Matrix[Num_Planes]; // filtered state vector
```

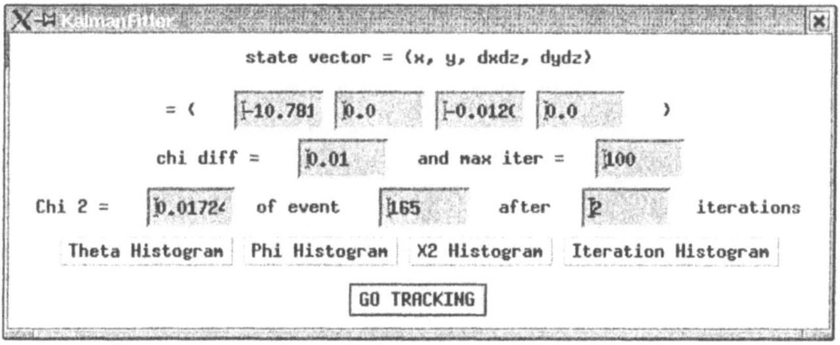

Figure 13.15. User interface for the Kalman tracker

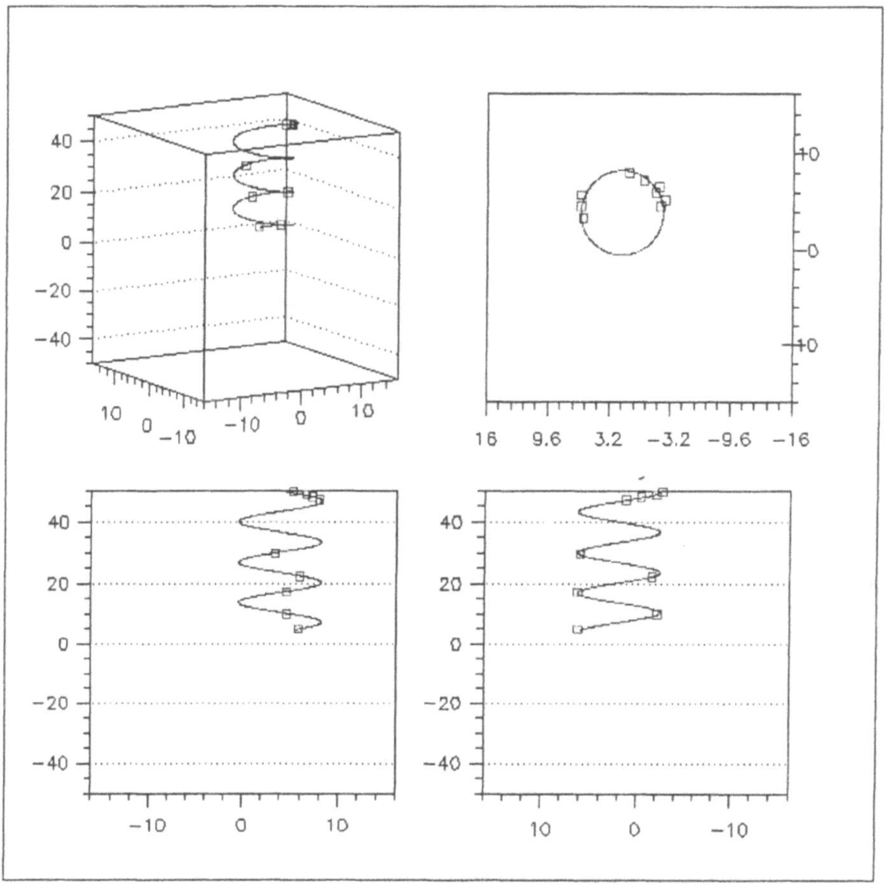

Figure 13.16. Helical track by Kalman method. Blue line is the best fit while red squares are recorded coordinates of hit wires.

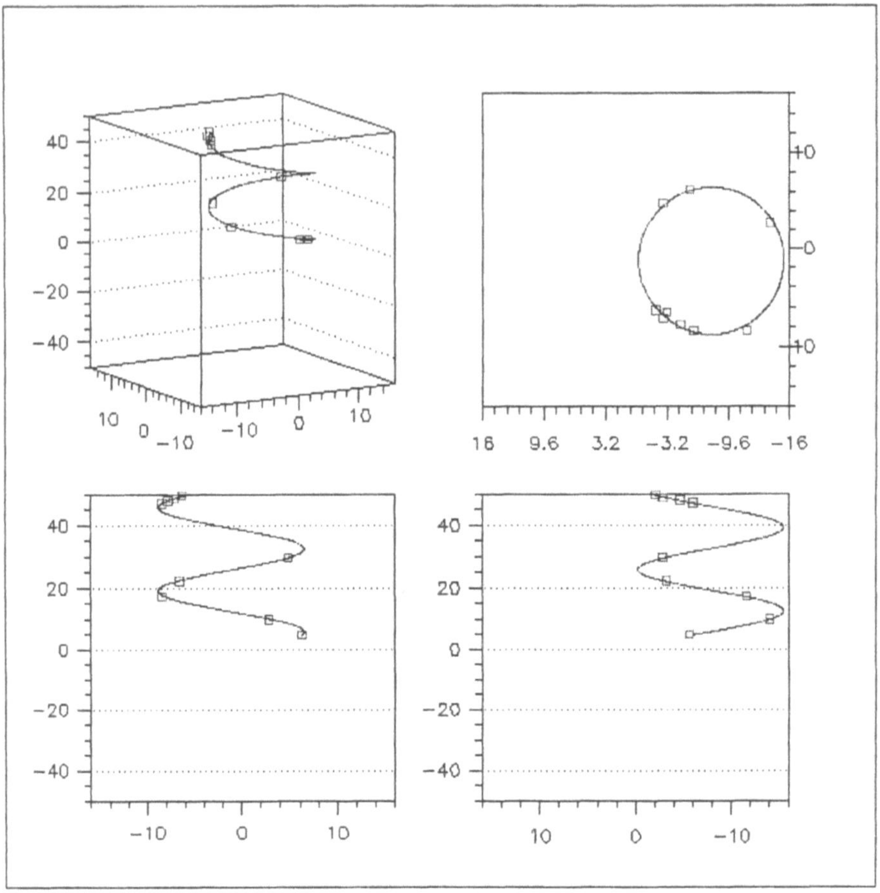

Figure 13.17. Helical track by Kalman method

```
Matrix[] Cf    = new Matrix[Num_Planes]; // filtered covariance matrix
Matrix[] F     = new Matrix[Num_Planes]; // transport matrix

Matrix[] xs    = new Matrix[Num_Planes]; // smoothed state vector
Matrix[] Cs    = new Matrix[Num_Planes]; // smoothed covariance matrix
Matrix[] Chi2s = new Matrix[Num_Planes]; // chi2 for smoothed state

KFDialog   parent;

public Kalman(KFDialog parent) {
       this.parent = parent;

       for (int i=0; i<Num_Planes; i++) {
           xp[i]    = new Matrix(5,1);
           Cp[i]    = new Matrix(5,5);
           xm[i]    = new Matrix(2,1);
           H[i]     = new Matrix(2,5);
           xf[i]    = new Matrix(5,1);
           Cf[i]    = new Matrix(5,5);
           F[i]     = new Matrix(5,5);
           xs[i]    = new Matrix(5,1);
           Cs[i]    = new Matrix(5,5);
           Chi2s[i] = new Matrix(1,1);
```

```
        }

        V   = new Matrix(2,2);
        rp  = new Matrix(2,1);
        G   = new Matrix(2,2);
        Kf  = new Matrix(5,2);
        rf  = new Matrix(2,1);
        As  = new Matrix(5,5);
        rs  = new Matrix(2,1);
        Rs  = new Matrix(2,2);

        tmp1= new Matrix(5,2);
        tmp2= new Matrix(5,5);
        tmp3= new Matrix(1,2);
        tmp4= new Matrix(1,2);
        reducedCp = new Matrix(4,4); // for straight lines
        inverseTmp= new Matrix(4,4);
        Cp_inverse= new Matrix(5,5);

        x = 0.0;
        y = 0.0;
        dxdz = 0.0;
        dydz = 0.0;
        epsilon = 0.1;
        iteration_max = 5;
        grandIteration = 0;

        z0 = new double[Num_Planes];
}       // end of Kalman class constructor

public double ChiSqr() {        // calculate the chi-square
        double sum;
        double[] d = new double[Num_Planes];
        double[] par = new double[4];
        double sigma = 0.2;

        // x = A*z + B
        // y = C*z + D

        par[0] = dxdz;
        par[1] = x;
        par[2] = dydz;
        par[3] = y;

        if (nHits[0] > 0) d[0]=(par[0]*z[0]+par[1]-xw[0]);
        if (nHits[1] > 0) d[1]=(par[2]*z[1]+par[3]-yw[0]);
        if (nHits[2] > 0) d[2]=(par[0]*z[2]+par[1]-xw[1]);
        if (nHits[3] > 0) d[3]=(par[2]*z[3]+par[3]-yw[1]);
        if (nHits[4] > 0) d[4]=(par[2]*z[4]+par[3]-yw[2]);
        if (nHits[5] > 0) d[5]=(par[0]*z[5]+par[1]-xw[2]);
        if (nHits[6] > 0) d[6]=(par[2]*z[6]+par[3]-yw[3]);
        if (nHits[7] > 0) d[7]=(par[0]*z[7]+par[1]-xw[3]);

        sum = 0.0;
        for (int j=0; j<Num_Planes; j++)
            if (nHits[j] > 0) sum += d[j]*d[j]/sigma/sigma;

        return sum;
}       // end of ChiSqr

public void Go() {
        int i;
        double x2old, x2diff;

        x = 0.0;
        y = 0.0;
        dxdz = 0.0;
```

```
        dydz = 0.0;

        x2old = ChiSqr();

        grandIteration = 0;
        i = 0;

        try {
            do {
              filter_smoother();
              X2 = ChiSqr();
              grandIteration += 1;
              i += 1;
              x2diff = Math.abs(x2old - X2);
              x2old = X2;
            } while (x2diff > 0.1 && i < 5);
        } catch (TrackerExceptions te) {
            X2 = Double.MAX_VALUE;
            System.out.println(te.getMessage());
        }
}

public void filter_smoother() throws TrackerExceptions {
        int    k, n, noneZeros;
        double Chi2Save, Chi2sDiff = 0.;
        double dz;
        double dtmp;

        // initialize
        k = 0;
        for (int j=0; j<Num_Planes/2; j++) {
            if (nHits[j] > 0) {
               n = j/2;
               if ((j%2) == 0) {  // x y x y
                  xm[k].M[0][0]  = xw[n];    // coordinate of
                  xm[k].M[1][0]  = 0.0;      // the fired wire
                  H[k].M[0][0]   = 1.0;
                  H[k].M[1][1]   = 0.0;
               } else {
                  xm[k].M[0][0]  = 0.0;
                  xm[k].M[1][0]  = yw[n];
                  H[k].M[0][0]   = 0.0;
                  H[k].M[1][1]   = 1.0;
               }
               z0[k]  = z[j];
               k += 1;
            }  // end if
        }
        for (int j=Num_Planes/2; j<Num_Planes; j++) {
            if (nHits[j] > 0) {
               n = j/2;
               if ((j%2) != 0) {  // y x y x
                  xm[k].M[0][0]  = xw[n];
                  xm[k].M[1][0]  = 0.0;
                  H[k].M[0][0]   = 1.0;
                  H[k].M[1][1]   = 0.0;
               } else {
                  xm[k].M[0][0]  = 0.0;
                  xm[k].M[1][0]  = yw[n];
                  H[k].M[0][0]   = 0.0;
                  H[k].M[1][1]   = 1.0;
               }
               z0[k]  = z[j];
               k += 1;
            }  // end if
        }
        noneZeros = k;
```

```
xp[0].M[0][0]  = x;          // x
xp[0].M[1][0]  = y;          // y
xp[0].M[2][0]  = dxdz;       // dxdz
xp[0].M[3][0]  = dydz;       // dydz
xp[0].M[4][0]  = 0.0;        // not used for straight tracks

V.M[0][0]      = 0.2*0.2;    // the wire spacing is 0.2 cm which
V.M[1][1]      = 0.2*0.2;    // defines the measurement error

Chi2Save       = 0.0;
iteration      = 0;

try {
 do {                                          // start iterating
   // start Kalman filter
   // Cp[0] is not critical for straight-line tracking
   dtmp = 12.0/((iteration+1)*10.0 + grandIteration*100.0) +
                              xm[0].M[0][0]-xp[0].M[0][0];
   Cp[0].M[0][0] = dtmp * dtmp;
   dtmp = 12.0/((iteration+1)*10.0 + grandIteration*100.0) +
                              xm[0].M[1][0]-xp[0].M[1][0];
   Cp[0].M[1][1] = dtmp * dtmp;
   dtmp = 12.0/((iteration+1)*10.0 + grandIteration*200.0);
   Cp[0].M[2][2] = dtmp * dtmp;
   dtmp = 12.0/((iteration+1)*14.0 + grandIteration*280.0);
   Cp[0].M[3][3] = dtmp * dtmp;

   Cp[0].M[4][4] = 0.0;

   for (k=0; k<noneZeros; k++) {

       rp.M = // residuals of predictions
       xm[k].minus(H[k].times(xp[k].M));

       G.M =
       V.plus(H[k].times(Cp[k].times(H[k].transpose())));

       tmp1.M =
       Cp[k].times(H[k].transpose());


       Kf.M = // Kalman gain matrix
       tmp1.times(G.ret_inv());

       xf[k].M = // Kalman prediction
       xp[k].plus(Kf.times(rp.M));

       tmp2.M =
       Kf.times(H[k].times(Cp[k].M));

       Cf[k].M = // update the covariance matrix
       Cp[k].minus(tmp2.M);

       rf.M = // filtered residuals
       xm[k].minus(H[k].times(xf[k].M));

       if (k < (noneZeros-1)) {
          dz = z0[k+1] - z0[k];
          TrackModel(k, dz);  // takes care of xp and Cp
          Cp[k+1].M = // extrapolation of the covariance
          F[k].times(Cf[k].times(F[k].transpose()));
          // note any process noise is to be added above
       }
   }   // end of filtering
```

```
            // start smoothing
            xs[noneZeros-1].M = xf[noneZeros-1].M;
            Cs[noneZeros-1].M = Cf[noneZeros-1].M;
            Chi2sTotal = 0.0;

            for (k=(noneZeros-1); k>0; k--) {

                for (int i=0; i<4; i++)
                    for (int j=0; j<4; j++)
                        reducedCp.M[i][j] = Cp[k].M[i][j];
                // invert a 4x4 instead of a 5x5 matrix
                inverseTmp.M = reducedCp.ret_inv();

                for (int i=0; i<4; i++)
                    for (int j=0; j<4; j++)
                        Cp_inverse.M[i][j] = inverseTmp.M[i][j];

                tmp2.M =
                Cf[k-1].times(F[k-1].transpose());

                As.M = // smoother gain matrix
                tmp2.times(Cp_inverse.M);

                xs[k-1].M = // smoothed state vector
                xf[k-1].plus(As.times(xs[k].minus(xp[k].M)));

                tmp2.M =
                Cs[k].minus(Cp[k].M);
                // covariance matrix of the smoothed state vector
                Cs[k-1].M =
                Cf[k-1].plus(As.times(tmp2.times(As.transpose())));

                rs.M = // smoothed residuals
                xm[k-1].minus(H[k-1].times(xs[k-1].M));

                Rs.M = // covariance matrix of smoothed residuals
                V.minus(H[k-1].times(Cs[k-1].times(
                                     H[k-1].transpose())));
                tmp3.M = rs.transpose();

                tmp4.M = tmp3.times(Rs.ret_inv());

                Chi2s[k-1].M = tmp4.times(rs.M);

                Chi2sTotal += Chi2s[k-1].M[0][0];
                Chi2sDiff = Math.abs(Chi2sTotal-Chi2Save);
                Chi2Save  = Chi2sTotal;

            }   // end of smoothing

            xp[0].M = xs[0].M;          // for next iteration
            iteration += 1;
         } while((Chi2sDiff>epsilon) && (iteration<iteration_max));
        } catch (TrackerExceptions te) {
         throw new TrackerExceptions(
                 "Matrix Inversion Error in Kalman");
        }
        x    = xp[0].M[0][0];       // x
        y    = xp[0].M[1][0];       // y
        dxdz = xp[0].M[2][0];       // dxdz
        dydz = xp[0].M[3][0];       // dydz
}       // end of method filter_smoother

// straight lines
public void TrackModel(int k, double dz) {
        xp[k+1].M[0][0] = xp[k].M[0][0] + xp[k].M[2][0]*dz;
        xp[k+1].M[1][0] = xp[k].M[1][0] + xp[k].M[3][0]*dz;
```

```
            xp[k+1].M[2][0] = xp[k].M[2][0];
            xp[k+1].M[3][0] = xp[k].M[3][0];

            F[k].M[0][0]    = 1.0;
            F[k].M[0][2]    = dz;
            F[k].M[1][1]    = 1.0;
            F[k].M[1][3]    = dz;
            F[k].M[2][2]    = 1.0;
            F[k].M[3][3]    = 1.0;
    }
}           // end of Kalman class
```

Listing 13.5 Kalman.java

Figures 13.5 to 13.14 show some straight-line tracking examples. The order of the wire planes is *x y x y y x y x* and different pairs are colored differently as seen in the figures. Black lines are results of the Kalman tracking (filtering and smoothing). During the analysis, results of the Kalman tracking are also returned to the dialog box event by event, as shown in Figure 13.15.

Figures 13.16 and 13.17 are examples of helices reconstructed by the Kalman method. Blue lines are the best estimates and red squares are the measured data. Note that the class to plot these helices is not part of the Java package enclosed in this chapter.

13.10 H Infinity Filter

Kalman filters work perfectly as long as the dynamics of the system is correctly modeled and distributions of the process disturbance and measurement error are Gaussian with zero means. What if the model is not precisely known and/or the disturbance/error is not Gaussian? Then Kalman filters are likely to fail. Consider the filter is used to track a missile or satellite. We can sacrifice a bit precision but we can never afford to lose tracking the missile. H_∞ (H infinity) filters are robust in that they return optimal estimates of the parameters given the worst-case process disturbance and measurement error in contrast to Kalman filters which estimate by minimizing the mean error variance.

We rewrite the system equation of Eq. (13.6) and measurement equation of Eq. (13.7) in the following form,

$$\begin{aligned} x_{k+1} &= F_k x_k + B_k u_k, \\ m_k &= H_k x_k + D_k v_k, \\ z_k &= L_k x_k, \end{aligned} \tag{13.31}$$

where x are the state vector, m the measurement vector, and z the vector to be estimated. u and v are respectively the process disturbance and measurement error, which are not required to be zero-mean Gaussian. k is index for (discrete) time. F_k, B_k, H_k, D_k, and L_k are matrices of appropriate dimensions at time k. We assume that $R_k \equiv D_k D_k^T > 0$ is true for any k.

Researchers most often find themselves playing games against nature. We want to find the estimate of $\hat{z}_k$ which minimizes $\sum_{i=0}^{N} \|z_k - \hat{z}_k\|^2$ while x_0, u_k, and v_k are doing the opposite, i.e., maximizing the squared estimate error. ($\hat{z}_k$'s are the true values.) Since the estimate error can be indefinitely large if $||u_k||, ||v_k||$, and the error in the initial state x_0 are arbitrarily large. We can never win. However, we define the cost function J to be minimized as follows,

$$J(\hat{z}; x_0, u, v) = \sum_{i=0}^{N} ||z_k - \hat{z}_k||^2 - \gamma^2 \cdot \left(\sum_{i=0}^{N} ||u_k||^2 + \sum_{i=0}^{N} ||v_k||^2 + ||x_0 - \bar{x}_0||^2_{C_0^{-1}} \right), \tag{13.32}$$

where γ is a positive definite constant. Since it multiplies with the disturbance and errors, it works as a penalty against arbitrarily large disturbance and errors. The so-called finite horizon, discrete-time, time-varying H_∞ filtering problem is to find estimates of $\hat{x}_k$ and $\hat{z}_k$ that satisfies,

$$max_{u,v,x_0} \frac{\sum_{i=0}^{N} ||z_k - \hat{z}_k||^2}{\sum_{i=0}^{N} ||u_k||^2 + \sum_{i=0}^{N} ||v_k||^2 + ||x_0 - \bar{x}_0||^2_{C_0^{-1}}} < \gamma^2. \tag{13.33}$$

For a very large value of γ, the second term of J dominates. The problem reduces to squared error minimization. It is therefore evident that the solution to the H_∞ filtering should approach to that of Kalman when γ tends to infinity.

The H_∞ filtering problem is more easily solved by a game theoretic approach. Moreover, the solution can fortunately be cast into an progressive form similar to that of Kalman. Here, we simply write down the solution which achieves the above H_∞ error bound.

$$\begin{aligned} x_k &= x_k^{k-1} + K_k(m_k - H_k x_k^{k-1}) \\ x_{k+1}^k &= F_k x_k \\ z_k &= L_k x_k \\ K_k &= C_k H_k^T (R_k + H_k C_k H_k^T)^{-1} \end{aligned} \tag{13.34}$$

where, starting with C_0 which is given a priori, C_k satisfies the so-called Raccati difference equation,

$$C_{k+1} = F_k C_k \big(I + (H_k^T R_k^{-1} H_k - \gamma^{-2} L_k^T L_k) C_k\big)^{-1} F_k^T + B_k B_k^T, \tag{13.35}$$

where I is an identity matrix of appropriate dimension. It is obvious that solutions exit only for certain values of γ. The conditions are,

$$\gamma^2 I - L_k C_k (I + H_k^T R_k^{-1} H_k C_k)^{-1} L_k^T > 0, \tag{13.36}$$

for all k.

13.11 Properties of H Infinity Filters

It can be shown that the magnitude of the gain matrix K_k of an H_∞ filter is always (i.e., for all k) larger than or equal to that of a Kalman. If we look at the first equation in Eqs. (13.34), the implication is that H_∞ filters are more sensitive to $(m_k - H_k x_k^{k-1})$. In other words, an H_∞ will converge to the actual state faster than a Kalman filter, given the same level of measurement uncertainty v_k.

γ enters into the C_k matrices and Eq. (13.36) is equivalent to saying that all the eigenvalues of the C_k's should have magnitudes less than one. γ, as a result, can not be set to too small a value.

A close investigation says that as γ gets large, C_k and the mean squared estimate error become small. On the other hand, the H_∞ becomes less sensitive to the measurement error v_k. Therefore, parameter γ represents a trade-off between mean estimate error and sensitivity to measurement error.

We have been stating that H_∞ filters are robust to unknown process disturbance and measurement error. Referring to the first two equations in Eqs. (13.31), we could, in the other perspective, have said that H_∞ filters are robust to unknown system models F_k and H_k.

In off-line analysis, we are interested in the best estimate of the state at any time, given all the measurements. This is namely the H_∞ smoother problem. The solution is identical to the Kalman smoother since in this case all the uncertainties have become known.

13.12 Summary

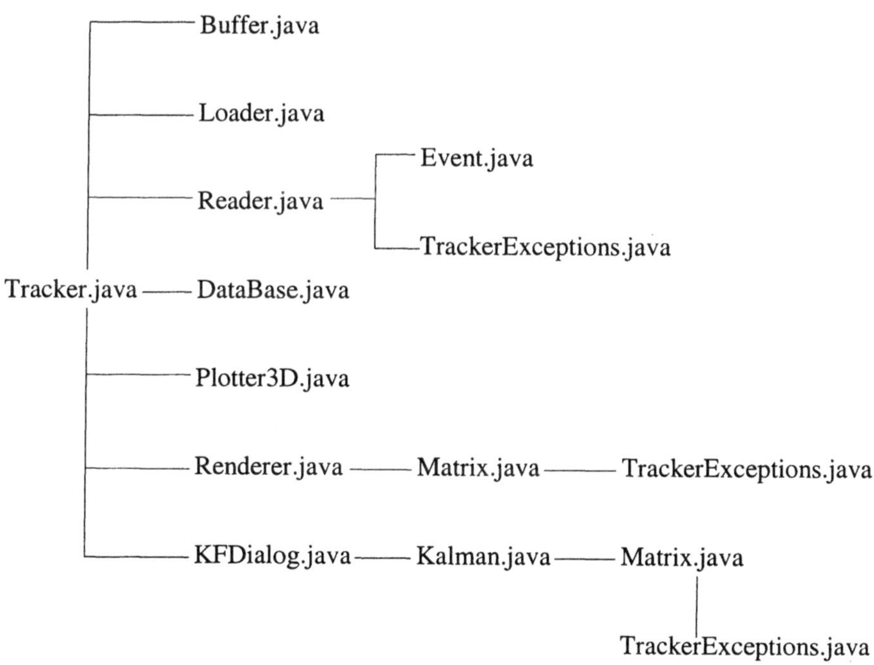

Figure 13.18. Interdependence of the source programs for the Kalman tracking

`TrackerExceptions.java`, `DataBase.java`, `Plotter3D.java`, `Renderer.java`, `KFDialog.java`, and `Tracker.java` are similar to corresponding classes in previous chapters. `Matrix.java` is the same class as seen in Chapter 1.

We introduced H_∞ filters which are applicable to unknown process disturbance and measurement error. It is also robust to uncertainties in system modeling.

We showed particle tracking using the Kalman algorithm. Tracks are predicted and filtered downstream. They are then smoothed backward to the upstream. The smoothed track parameters use information from all hits and are thus the best-fit parameters. The Kalman method is most useful in real-time applications.

Kalman filter is a recursive procedure, where the prediction from the propagation model is updated once new measurement is available. The weighting in the prediction and updating balances between process noise and measurement error. If the error and noise are Gaussian distributed, the method is proved to be equivalent to maximum likelihood methods such as chi-square fits.

Kalman filter is a subset of graphical model whose structure is known and fixed. In more general graphical modeling, both structure and parameters in the structure are unknown and varying. Probability distributions are used to mediate between graphical models and the real world.

13.13 References and Further Reading

Bayesian information criterion was derived in, G. Schwartz, "Estimating the dimension of a model", *Ann. Statist.*, **6** (1978) 461-464. Akaike information criterion was due to, H. Akaike, "A new look at the statistical model identification", *IEEE Trans. Automat. Contr.*, AC-**19** (1974) 716-723

The Kalman method was first proposed in, R.E. Kalman, "A New Approach to Linear Filtering and Predicting Problems", *Transaction of the ASME-Journal of Basic Engineering*, **82** (Series D) (1960) 35-45

For the Kalman formulas, we have followed the notation of, R. Fruhwirth, "Application of Kalman Filtering to Track and Vertex Fitting", *Nucl. Instr. & Methods*, **A262** (1987) 444-450

A tutorial derivation of the Kalman filter can be found in the downloadable article, A.L. Barker, D.E. Brown, and W.N. Martin, "Bayesian Estimation and the Kalman Filter", *IPC-TR-94-002*, Revised Sept. 19, 1994. Its published version appears in, *Computers and Mathematics with Applications*, Vol. 30, No. 10, 1995

A popular Kalman web site is `http://www.cs.unc.edu/~welch/kalman`, which contains and refers to tutorials, books, downloadable articles, researches, links, as well as software.

A textbook on optimal filtering is, B.D.O. Anderson and J.B. Moore, "Optimal Filtering", Prentice-Hall, Englewood Cliffs, N.J. (1979)

Solutions to the H infinity problem are shown in, K.M. Nagpal and P.P. Khargonekar, "Filtering and Smoothing in an H_∞ Setting", *IEEE Trans. Automat. Contr.*, AC-**36**, (1991) 152-166

A review of H infinity filters is, U. Shaked and Y. Theodor, "H_∞-Optimal Estimation: A Tutorial", *Proceedings of the 31st Conference on Decision and Control, Tucson, Arizona, December 1992*, pp. 2278-2286

A useful thesis with examples on H infinity is, K. Takaba, "Studies on H_∞ Filtering Problems for Linear Discrete-Time Systems", doctoral dissertation in Applied Mathematics and Physics, Kyoto University, January 1996

Chapter 14

JNI TECHNOLOGY

Before the introduction of Java programming language, there had existed many collections of general-purpose or specialized subroutines or functions written in other languages, such as Fortran and C. These well-established and tested libraries represent decades' efforts of experts in an assortment of fields. As a programmer in Java, she might want to reap the fruits instead of redoing the hard jobs. The capability of calling methods in other languages from within Java programs is accomplished by Java Native Interface (JNI). We provide an example of calling CERN (European Organization for Nuclear Research) library routines in this chapter.

14.1 Java Native Interface

Most institutions provide their employees with computing facilities, including hardware platforms and software. The server machines of the institutions are most often installed with various software packages to suit users' needs. Many of the packages are written in languages other than Java. As end users, what we can do to take advantage of the packages is to call the native methods (methods written in languages other than Java) via *Java Native Interface* technology.

We will show an example of calling Fortran routines from within a Java program. The example Java program firstly calls a C program, which in turn calls a Fortran routine, which in turn calls the designated routine in a shared Fortran library. Through the exercise, it is hoped that most frequently encountered basic issues regarding inter-language programming are addressed.

14.2 JNI HOW-TO

We outline the general steps to run native methods (methods in C in this section) from a Java program. First of all, the method in C is declared as `native` in the field of the class which is invoking the native method. The native library is then loaded by the statement in the native method calling class,

```
System.loadLibrary("MyNativeC");
```

where `MyNativeC` (to be created later) is a shared object library (i.e., libMy-NativeC.so). The native method can now be called by the way it is supposed to. That's all we have to do with the Java source.

The second step is to compile the Java source,

```
$javac MyNative.java
```

where $ is the system prompt, and `MyNative.java` contains the native method.

Step 3 is to generate a header file for the C program,

```
$javah -jni MyNative
```

after which `MyNative.h` is created.

The job left is to write the C program. The header file, created in step 3, tells us how to declare the method in C. The tough part is argument passing to/from the C method. JNI provides standard methods to handle primitive data types, and their arrays. For example, to pass strings, method `GetStringUTFChar()` is used for conversion because a string in Java is a a 16-bit Unicode, whereas a 8-bit character in C.

After the C program is finished, the final step is to compile it (which we assume is `mynative.c`) with the C compiler,

```
$gcc -o libMyNativeC.so -shared mynative.c
```

in a GNU C/Linux platform. This is all if we call C programs from Java. (For Fortran methods, the above compilation command is a little bit different.) To run the program, do what is normally done for Java applications, namely, `java MyNative`.

To proceed with our more involved example of Fortran native methods, we need to know how to call Fortran from C.

14.3 Call Fortran Programs from C

The right procedure is highly machine dependent though the way introduced in this section applies to most platforms.

First of all, the Fortran routine (or function) `sub` is declared `extern sub_` in the C program. Note the lower case and the appended underscore. This convention depends on the Fortran compiler of the host machine. Refer to user's manual in case it does not work.

Fortran `COMMON` blocks can be accessed from C code via C structures. For example,

```
c
      integer i
      real x
      double precision d
      common / mycomx / i,x(3,3),d

      character*80 ctext(10)
      common / mycomc / ctext
c

// The above blocks can be accessed through C structures
extern struct {
       int i;
       float x[3][3];
       double d;
} mycomx_;

extern struct {
       char ctext[10][80];
} mycomc_;
```

Note again the appended underscores and that x[1][2] in C refers to x(3,2) in Fortran because C stores array elements row-wise, while Fortran column-wise. It is also reminded that array elements start from 0 in C as in Java, while 1 in Fortran.

We now face the same mess of argument passing between C and Fortran as between Java and C. Arguments in Fortran are all passed by *reference*. We therefore have to prepare the corresponding pointers in C before passing them. The more trouble part is with character variables. Again, this is machine dependent. The way described here is applicable to most but not all Fortran compilers. Besides the pointer to the character string, we need to specify the length of the string, and this additional information (namely an `int`) is placed in the end of the argument list of the Fortran routine. The following example illustrates it,

```
// in order to call Fortran routine
//
//        sub(x,ch_ptr,y),
//        real x,y
//        character*(*) ch_ptr
//
// we need to do the following in C

float x,y;
char* ch_ptr;
int   ch_len;
```

```
sub_(&x,ch_ptr,&y,ch_len);
```

14.4 A JNI Example

The goal of this example is to call from Java a routine in a Fortran library which performs curve fitting. The Fortran routine, called `HFITV()`, is a standard multi-dimensional histogram fitting routine in CERN library. As mentioned earlier, we will achieve this Java/Fortran link via an intermediary C program. Therefore, in the Java program, called `Hfitv.java`, data are read into arrays from a file. The data arrays are then passed to the intermediary C program, in which the Fortran HFITV() is called. Figure 14.1 shows the involved files and also the JNI steps. The data file, `wireT0_934.dat`, contains 80×8 histograms each of which looks like the one plotted in Figure 14.2. In the histogram, the counts up to the peak are fitted to an exponential function while counts from the peak to the rightmost bin are fitted to an *error function.* This fit is done in batch by the CERN library routine `HFITV()`. The best-fit parameters are returned to the Java calling class which also outputs them on the terminal screen.

For clarity, we split the jobs in Java into two classes. Class `Hfitv` (Listing 14.1), with method `main()`, copes with reading data from files. The second Java class, Hbook, contains the native method declaration and shared library loading (Listing 14.2).

```
/*      Sun-Chong Wang
        TRIUMF
        4004 Wesbrook Mall
        Vancouver, V6T 2A3
        Canada
        e-mail: wangsc@triumf.ca

        Hfitv.java reads data and calls the native method which fits
        the data to the Fortran function timeshape(x) in myhfitv.f */

import java.io.*;

class Hfitv {
 public static void main(String args[]) {
        double tdc[] = new double[60];
        double plane[][] = new double[8][60];
        double count[] = new double[60];
        double countErr[] = new double[60];
        double par[] = new double[5];
        double chi2 = 0.0f;
        Hbook hbook = new Hbook();

        // read data from file
        System.out.println("Reading "+args[0]);
        try {
             FileInputStream istream = new FileInputStream(args[0]);
             InputStreamReader br     = new InputStreamReader(istream);
             BufferedReader    re     = new BufferedReader(br);
             StreamTokenizer   sto    = new StreamTokenizer(re);

             par[0] = 300.0f;  // these are initial guess
             par[1] = 5925.0f;
```

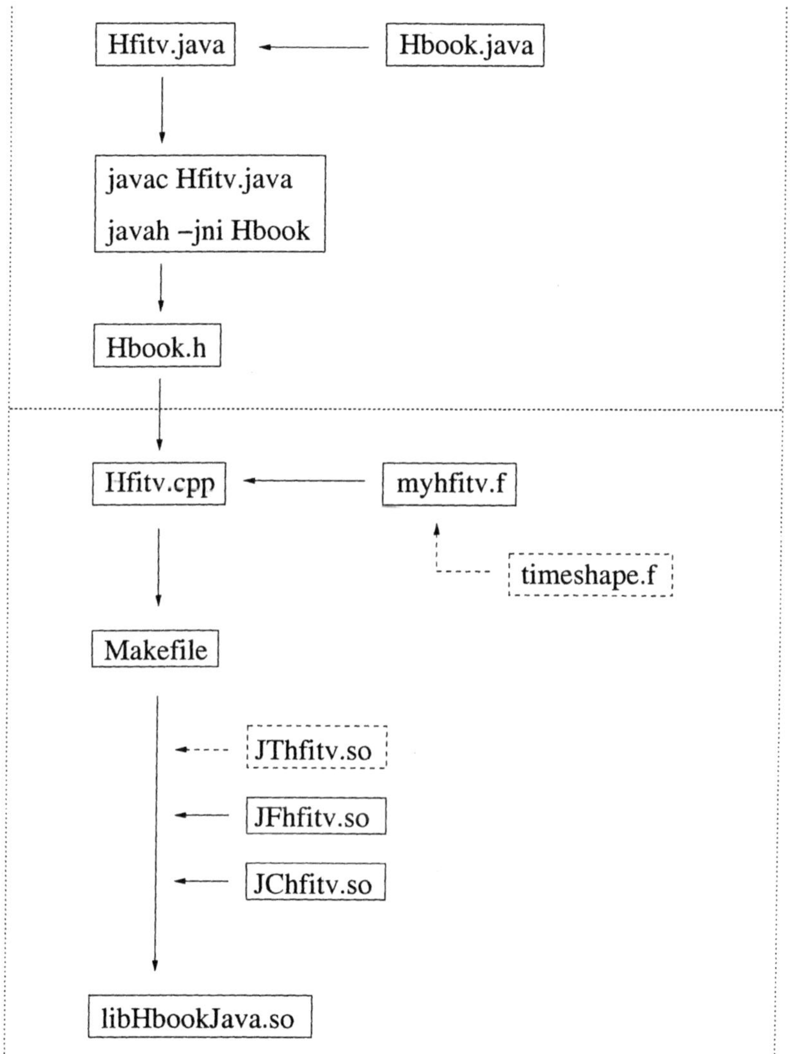

Figure 14.1. Files and flowchart of the JNI steps

```
par[2] = 17.0f;
par[3] = 5965.0f;
par[4] = 50.0f;

// how to loop has to do with the data format in the file
for (int i=0; i<80; i++) {
    for (int j=0; j<60; j++) {
        sto.nextToken();
        sto.nextToken();
        tdc[j] = sto.nval;
        for (int k=0; k<8; k++) {
```

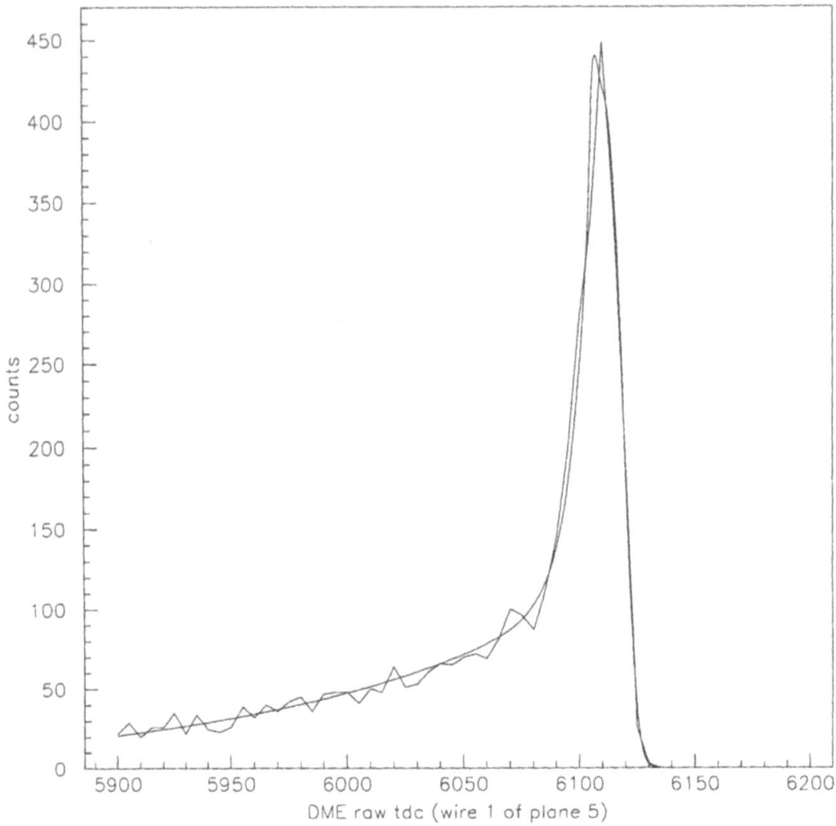

Figure 14.2. Piece-wise lines connect the data while the smooth curve results from the fit.

```
                        sto.nextToken();
                        plane[k][j] = sto.nval;
                    }
                }
            }

            for (int i=0; i<80; i++) {
                for (int j=0; j<8; j++) {
                    for (int k=0; k<60; k++) {
                        count[k] = plane[j][k];
                        countErr[k] = Math.sqrt(count[k]);
                    }
                    // send the x, y, and error in y for fits
                    hbook.HfitvSetup(tdc,count,countErr,par,chi2);
                    hbook.HfitvGo();
                    // print the resulting best-fit parameters
                    System.out.println(j+" "+i+" "+par[0]+" "+par[1]
                               +" "+par[2]+" "+par[3]+" "+par[4]);
                }
            }
            istream.close();
        } catch (IOException e) {
```

```
                System.out.println("Fail: "+e.getMessage());
        }
  }     // end of main
}       // end of class
```

Listing 14.1 Hfitv.java

```
/*      Sun-Chong Wang
        TRIUMF
        4004 Wesbrook Mall
        Vancouver, V6T 2A3
        Canada
        e-mail: wangsc@triumf.ca

        Hbook.java installs the native method */

class Hbook {
 int nrecl = 1024;                // this appeared magic in my system
 int nchan, ndim, ndis, npar;     // required by the native method
 double[] tdc, count, countErr;   // x, y, and the errors
 double[] par, sigpar;            // parameters and the uncertainties
 double chi2;                     // returned chi-square
 double[] dum1, dum2, dum3;       // required by the native method

 public Hbook() {
        super();
 }      // default constructor

 public void HfitvSetup(double[] tdc, double[] count, double[]
                        countErr, double[] par, double chi2) {
        this.tdc      = tdc;
        this.count    = count;
        this.countErr = countErr;
        this.par      = par;
        this.sigpar   = sigpar;
        nchan         = tdc.length;
        ndim          = nchan;
        ndis          = 1;
        npar          = par.length;
        dum1          = new double[npar];
        dum2          = new double[npar];
        dum3          = new double[npar];
        sigpar        = new double[npar];
        for (int i=0; i<5; i++) dum1[i] = -1.0; // dummy
 }

 public void HfitvGo() {
        myhfitv(nchan, ndim, ndis, tdc, count, countErr,
                npar, par, dum1, dum2, dum3, sigpar, chi2);
 }

 public native void myhfitv(int nchan, int ndim, int ndis,
                            double[] tdc, double[] count,
                            double[] countErr, int npar, double[] par,
                            double[] dum1, double[] dum2, double[] dum3,
                            double[] sigpar, double chi2);

 // Load DLL (or shared library) which contains implementation of
 // native methods
 static {
        System.loadLibrary("HbookJava");
 }
}       // end of class
```

Listing 14.2 Hbook.java

Having written the Java programs, we type, under the system prompt,

```
$javac Hfitv.java
```

and get `Hfitv.class` and `Hbook.class`. Then we issue,

```
$javah -jni Hbook
```

to get the header file `Hbook.h`, which is shown in Listing 14.3.

```
/* DO NOT EDIT THIS FILE - it is machine generated */
#include <jni.h>
/* Header for class Hbook */

#ifndef _Included_Hbook
#define _Included_Hbook
#ifdef __cplusplus
extern "C" {
#endif
/*
 * Class:     Hbook
 * Method:    myhfitv
 * Signature: (III[D[D[DI[D[D[D[D[DD)V
 */
JNIEXPORT void JNICALL Java_Hbook_myhfitv
  (JNIEnv *, jobject, jint, jint, jint, jdoubleArray, jdoubleArray,
   jdoubleArray, jint, jdoubleArray, jdoubleArray, jdoubleArray,
   jdoubleArray, jdoubleArray, jdouble);

#ifdef __cplusplus
}
#endif
#endif
```

Listing 14.3 Hbook.h

In the C++ (in fact a C) program, `Hfitv.cpp`, we define, with the guidance of `Hbook.h`, the C function and the corresponding arrays, as shown in Listing 14.4.

```
#include "Hbook.h"

#ifdef __FORTRAN_BUILD__

extern "C" void myhfitv_(long *nchan, long *ndim, long *ndis,
           double *ttdc, double *tcount, double *tcountErr,
           long *npar, double *tpar, double *tdum1, double *tdum2,
           double *tdum3, double *tsigpar, double *chi2);

JNIEXPORT void JNICALL Java_Hbook_myhfitv
(JNIEnv *env, jobject, jint nchan, jint ndim, jint ndis,
   jdoubleArray tdc, jdoubleArray count, jdoubleArray countErr,
   jint npar, jdoubleArray par, jdoubleArray dum1, jdoubleArray dum2,
   jdoubleArray dum3, jdoubleArray sigpar, jdouble chi2) {

        jdouble* ttdc = env->GetDoubleArrayElements(tdc,0);
        jdouble* tcount = env->GetDoubleArrayElements(count,0);
        jdouble* tcountErr = env->GetDoubleArrayElements(countErr,0);
        jdouble* tpar = env->GetDoubleArrayElements(par,0);
        jdouble* tdum1 = env->GetDoubleArrayElements(dum1,0);
        jdouble* tdum2 = env->GetDoubleArrayElements(dum2,0);
        jdouble* tdum3 = env->GetDoubleArrayElements(dum3,0);
        jdouble* tsigpar = env->GetDoubleArrayElements(sigpar,0);
```

```
        myhfitv_(&nchan,&ndim,&ndis,ttdc,tcount,tcountErr,
                 &npar,tpar,tdum1,tdum2,tdum3,tsigpar,&chi2);

        env->ReleaseDoubleArrayElements(tdc,       ttdc,0);
        env->ReleaseDoubleArrayElements(count,     tcount,0);
        env->ReleaseDoubleArrayElements(countErr,  tcountErr,0);
        env->ReleaseDoubleArrayElements(par,       tpar,0);
        env->ReleaseDoubleArrayElements(dum1,      tdum1,0);
        env->ReleaseDoubleArrayElements(dum2,      tdum2,0);
        env->ReleaseDoubleArrayElements(dum3,      tdum3,0);
        env->ReleaseDoubleArrayElements(sigpar,    tsigpar,0);
};
#endif
```

Listing 14.4 Hfitv.cpp

In the Fortran program, `myhfitv.f` (Listing 14.5), some arguments, required by `HFITV()`, are initialized. We do not bother passing them from `Hfitv.java` because they are never changed.

```
      subroutine myhfitv(nchan, ndim, ndis, tdc, count,
     +                   countErr, npar, par, dum1, dum2,
     +                   dum3, sigpar, chi2)
      integer i
      integer*4 nchan, ndim, ndis, npar
      real*8 chi2
      real*8 tdc(*), count(*), countErr(*)
      real*8 dum1(*), dum2(*), dum3(*)
      real*8 sigpar(*)
      real*8 par(*)

      integer hnchan, hndim, hndis, hnpar
      real hchi2
      real htdc(nchan), hcount(nchan), hcountErr(nchan)
      real hdum1(npar), hdum2(npar), hdum3(npar)
      real hsigpar(npar)
      common/fcb/par2(5)
      external timeshape

      hnchan = nchan
      hndim  = ndim
      hndis  = ndis
      hnpar  = npar

      do 10 i=1,nchan
         htdc(i) = tdc(i)
         hcount(i) = count(i)
         hcountErr(i) = countErr(i)
10    continue

      do 20 i=1,npar
         par2(i) = par(i)
         hdum1(i) = dum1(i)
         hdum2(i) = dum2(i)
         hdum3(i) = dum3(i)
         hsigpar(i) = sigpar(i)
20    continue

      CALL HLIMIT(1024)

      CALL HFITV(hnchan,hndim,hndis,htdc,hcount,hcountErr,
     +           timeshape,'Q',hnpar,par2,hdum1,hdum2,
     +           hdum3,hsigpar,hchi2)

      do 30 i=1,5
         par(i) = par2(i)
```

```
30    continue
      chi2 = hchi2
      write(*,*) par(1), par(2), par(3), par(4), par(5), hchi2

      return
      end

      real*8 function timeshape(x)
      real x
      common/fcb/par2(5)

      if (x .ge. par2(2)) then
         timeshape=par2(1)*0.5*erfc((x-par2(4))/par2(3)/1.4142)
      else
         timeshape=par2(1)*exp((x-par2(2))/par2(5))
      endif

      end
```

Listing 14.5 myhfitv.f

Finally, we provide the `Makefile` (Listing 14.6) which generates `JCHfitv.so` and `JFHfitv.so`. The former results from `gcc` compiling `Hfitv.cpp`, and the latter from `g77 myhfitv.f`. These two shared libraries are combined with the CERN library in the host machine to generate the final shared library, `libHbookJava.so`, which is to be used by the Java virtual machine when the native method is called. The reason for the last linking is that routine `HFITV()` and the error function, `erfc()`, reside in the CERN library.

The dashed box and line in Figure 14.1 mean optional. There arise occasions where we want to change the fitting function (the curve in Figure 14.2), which is `timeshape(x)` in our Fortran routine `myhfitv.f`. The provision of the fitting function in a separate program `timeshape.f` makes the change convenient.

```
CCFLAGS     = -shared
CPPDEFINES = -D __FORTRAN_BUILD__
FFLAGS      = -O -c

IDIRS   = -I/home/wangsc/JAVA/jdk1.2.2/include \
          -I/home/wangsc/JAVA/jdk1.2.2/include/linux

FFILES = JFhfitv.so
CFILES = JChfitv.so

libHbookJava.so : $(FFILES) $(CFILES)
        g77 $(CCFLAGS) -o libHbookJava.so $(FFILES) $(CFILES) \
'/triumfcs/linux/cern/i386_redhat61/pro/bin/cernlib packlib,mathlib'

JChfitv.so    : Hfitv.cpp
        gcc $(CCFLAGS) -o JChfitv.so $(CPPDEFINES) $(IDIRS) Hfitv.cpp

JFhfitv.so    : myhfitv.f
        g77 $(FFLAGS) myhfitv.f -o JFhfitv.so
```

Listing 14.6 Makefile

Note there is a tag before `gcc` (and `g77`) in the Makefile. Listing 14.7 shows part of the program output on the screen. Note that parameter values are also written out in the Fortran program, `myhfitv.f`, to double check with the values returned in Java.

```
Reading wireT0_934.dat
0 0 181.7764129638672 5916.89013671875 24.19029426574707 5949.1625976562
```

```
5 63.70079803466797
1 0 169.46859741210938 5928.951171875 16.067157745361328 5967.4174804687
5 53.52640914916992
2 0 142.6890106201172 5932.39990234375 18.925033569335938 5963.538085937
5 59.30392837524414
3 0 175.21937561035156 5925.85888671875 18.811729431152344 5965.93994140
625 60.03959274291992
4 0 166.66250610351562 5926.72705078125 19.18109703063965 5973.790039062
5 62.82880783081055
5 0 156.78761291503906 5951.49462890625 25.139509201049805 5963.66552734
375 83.39823150634766
6 0 211.97471618652344 5946.5908203125 23.05830955505371 5969.5927734375
 62.75101089477539
7 0 177.5975341796875 5948.14599609375 18.420970916748047 5962.217285156
25 56.40188980102539
0 1 181.62709045410156 5917.7392578125 23.351112365722656 5949.053222656
25 64.03817749023438
1 1 183.4449462890625 5934.90234375 17.715688705444336 5964.78369140625
53.55690002441406
...
 MINUIT RELEASE 96.03  INITIALIZED.   DIMENSIONS 100/ 50  EPSMAC=  0.89
E-15
 **********
 **    1 **SET EPS  0.1000E-06
 **********
 FLOATING-POINT NUMBERS ASSUMED ACCURATE TO   0.100E-06
 **********
 **    2 **SET ERR   1.000
 **********
  181.776413  5916.89014  24.1902943  5949.1626  63.700798  1.36133766
  169.468597  5928.95117  16.0671577  5967.41748  53.5264091  1.29428756
  142.689011  5932.3999  18.9250336  5963.53809  59.3039284  1.03486609
  175.219376  5925.85889  18.8117294  5965.93994  60.0395927  1.40554309
  166.662506  5926.72705  19.181097  5973.79004  62.8288078  1.31859028
  156.787613  5951.49463  25.1395092  5963.66553  83.3982315  1.24006283
  211.974716  5946.59082  23.0583096  5969.59277  62.7510109  1.43256497
  177.597534  5948.146  18.4209709  5962.21729  56.4018898  1.71147847
  181.62709  5917.73926  23.3511124  5949.05322  64.0381775  1.3697865
  183.444946  5934.90234  17.7156887  5964.78369  53.5569  1.31286204
...
```

Listing 14.7 Output on screen from the Java program

14.5 Summary

We showed a JNI example where Fortran library routines are invoked in Java via an intermediary C program. The example was meant to be comprehensive so that most JNI tasks can be done in a similar fashion.

An intermediary C program makes native calls flexible since C and C++ programs are widely used. In fact, many Fortran programs have C ports.

JNI proves a valuable tool. Java programmers, with the JNI technology, can make use of legacy utility routines, such as LAPACK, IMSL, NAG, or database packages written in other major languages.

14.6 References and Further Reading

Sun Microsystems' website has detailed information about how to map C arrays and C++ objects, and how to access class fields, and so on. Tutorials of JNI are available at `java.sun.com/docs/books/tutorial/native1.1/index.html`.

The following article contains a list of platform specific information on interfacing C and Fortran. A. Nathaniel, "6.1 Interface Fortran and C", *CERN Computer Newsletter*, **217** (1994) 9-15

CERN is where the World Wide Web was born. CERN program library information can be found online at `http://wwwinfo.cern.ch/asd`

Appendix A

A.1 Web Computing

The SETI institute launched a campaign harnessing the power of hundreds of thousands of Internet connected computers to analyze radio signals from space in search for extraterrestrial intelligence. A participant downloads data from SETI and the analysis runs as a screen saver program on the volunteer's machine at home.[1] Since then, many organizations have followed the model to attack computationally intensive jobs such as genome research.

To achieve such a web-based computing in Java, we show how easily a standalone application, such as the example program we wrote in each chapter of the book, can be converted into an applet which runs on the browser machine once the web page containing the applet is clicked. What happens is that the bytecodes of the applet class is downloaded from the web server to the browser machine and an instance of the class is created in the (Java-enabled) Web browser.[2] An applet, a subclass of `Panel`, contains, instead of the `main()` method, the `start()` method which is run on the browser machine. The constructor of an applet is the `init()` method. For security reasons, an applet is not allowed to write output on the file system of the browser machine. For other restrictions and methods of an applet, you are referred to Java's online manual at Sun Microsystems' website. Figure A.1 shows the web page containing the link to the applet. Once the 'here' is clicked, the applet is downloaded and the viewer can choose items in the window, changing parameters and starting program execution as shown in Figure A.2. Listing A.1 shows the html source for the web page in Figure A.2. Listing A.2 is the code to create the applet from the standalone application (lattice gas automata in this case). Note that there are now six (cf. Figure 9.38) class files for the applet. We've commented out the `main()` method in the `Hydro.java`. After the command `javac HydroApplet.java` at the system prompt, we archive the six classes into one using Java's `jar` utility,

```
$jar cvf ca.jar DataBase.class Hydro.class HydroApplet.class LGA.class L
 GADialog.class Plotter.class
```

[1]Details of the *SETI@home* project can be found at `http://setiathome.ssl.berkeley.edu`.

[2]Java Runtime Environment can be downloaded and plugged into the browser to make it applet savvy.

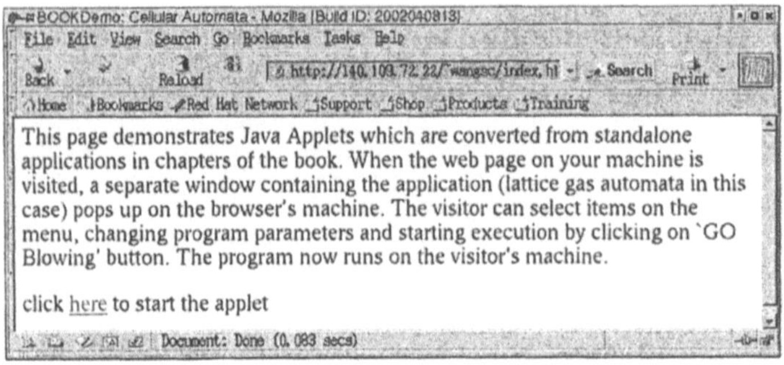

Figure A.1. Web page for the viewer to click on the applet

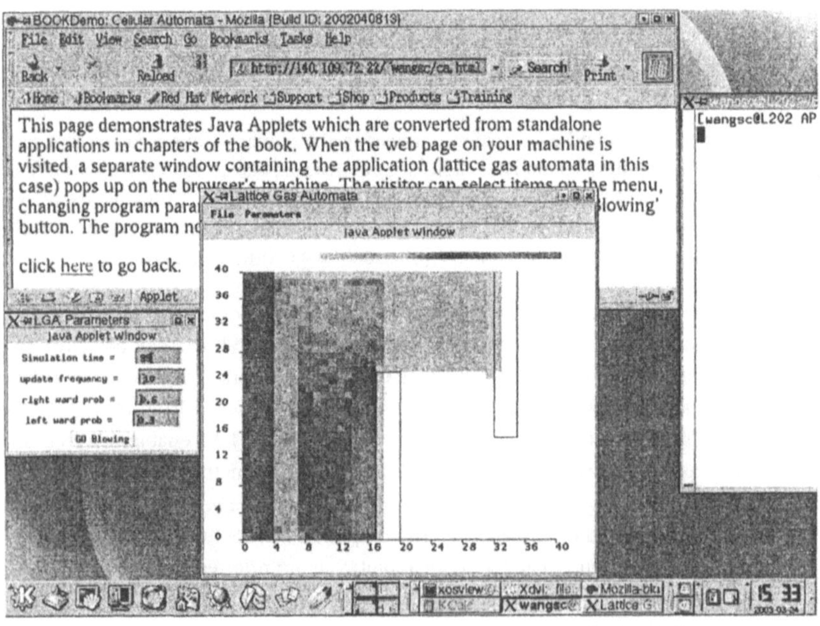

Figure A.2. Screen shot after the applet is loaded and executed

A file called `ca.jar` is then created in the same directory containing the lattice gas automata classes. We then move `ca.jar` to the subdirectory of our html files, `public_html/CA/`, as prescribed by the `codebase` attribute of the `applet` tag in the html file.

```
<HTML>
<HEAD>
<TITLE>BOOKDemo: Cellular Automata</TITLE>
</HEAD>

<BODY>
<p>
```

```
This page demonstrates Java Applets which are converted from standalone
applications in chapters of the book. When the web page on your machine
is visited, a separate window containing the application (lattice gas
automata in this case) pops up on the browser's machine. The visitor can
select items on the menu, changing program parameters and starting
execution by clicking on 'GO Blowing' button. The program now runs on
the visitor's machine.
</p>
click <a href="http://140.109.72.22/~wangsc/index.html">here</a> to go
back.
<applet code="HydroApplet.class" archive="ca.jar" codebase="CA/"
width=420 height=0>
</applet>
</BODY>
</HTML>
```

Listing A.1 ca.html

```
/*      Sun-Chong Wang
        TRIUMF
        4004 Wesbrook Mall
        Vancouver, V6T 2A3
        Canada
        e-mail: wangsc@triumf.ca

        HydroApplet.java creates a lattice gas automata applet */

import java.applet.Applet;

public class HydroApplet extends Applet {
 Hydro demo;
 public void start() {
        demo = new Hydro();
        demo.setVisible(true);
 }
}       // end of applet class
```

Listing A.2 HydroApplet.java

A.2 Class Sources

This section shows source codes to complete the applications in Chapters 2 and 4.

```
/*      Sun-Chong Wang
        TRIUMF
        4004 Wesbrook Mall
        Vancouver, B.C. V6T 2A3
        Canada
        e-mail: wangsc@triumf.ca

        FDialog.java creates a dialogue box accepting data format
        of the input file */

import java.lang.*;
import java.awt.*;
import java.awt.event.*;

class FDialog extends Dialog implements ActionListener {
 TextField nheadTF, colnxTF, ncolnTF;
 Integer nheadI, colnxI, ncolnI;
```

```
int nhead, colnx, ncoln;

MyWindow parent;

public FDialog(MyWindow parent, String text) {
        super(parent,text,true);
        setBackground(Color.white);
        this.parent = parent;

        setLayout(new GridLayout(4,1));

        // default format
        nhead = 4;      // no. of header lines to skip
        colnx = 1;      // x column
        ncoln = 5;      // total no. of columns

        Panel nheadP = new Panel();
        Panel colnxP = new Panel();
        Panel ncolnP = new Panel();

        nheadI = new Integer(nhead);
        colnxI = new Integer(colnx);
        ncolnI = new Integer(ncoln);

        nheadTF = new TextField(nheadI.toString(),6);
        colnxTF = new TextField(colnxI.toString(),6);
        ncolnTF = new TextField(ncolnI.toString(),6);

        nheadP.add(new Label("# of header lines to ignore:",
                                                Label.LEFT));
        nheadP.add(nheadTF);
        add(nheadP);

        ncolnP.add(new Label("Total # of columns:          ",
                                        Label.LEFT));
        ncolnP.add(ncolnTF);
        add(ncolnP);

        colnxP.add(new Label("Column ",Label.LEFT));
        colnxP.add(colnxTF);
        colnxP.add(new Label(" is x coordinates",Label.LEFT));
        add(colnxP);

        Panel bbP = new Panel();
        Button button1 = new Button("  OK      ");
        Button button2 = new Button(" Cancel  ");
        button1.addActionListener(this);
        button2.addActionListener(this);

        bbP.add(button1);
        bbP.add(button2);

        add(bbP);
        setSize(new Dimension(320,150));
}

// action handler
public void actionPerformed(ActionEvent e) {
        if ("  OK      ".equals(e.getActionCommand())) {
           nhead = nheadI.parseInt(nheadTF.getText());
           ncoln = ncolnI.parseInt(ncolnTF.getText());
           colnx = ncolnI.parseInt(colnxTF.getText());
           parent.nSkips = nhead;
           parent.nColns = ncoln;
           parent.xIndex = colnx;
           dispose();
        }
        if (" Cancel  ".equals(e.getActionCommand())) {
```

```
            dispose();
        }
 }      // end of method actionPerformed
}       // end of FDialog Class
```

Listing A.3 FDialog.java

```
/*      Sun-Chong Wang
        TRIUMF
        4004 Wesbrook Mall
        Vancouver, B.C. V6T 2A3
        Canada
        e-mail: wangsc@triumf.ca

        Message.java puts a message box on the screen indicating
        work is in progress */

import java.awt.*;

class Message extends Dialog {
 public Message(Frame parent, String title, String message) {
        super(parent,message,false);
        setBackground(Color.white);
        setLayout(new GridLayout(1,1));
        add(new Label(title,Label.LEFT));
        pack();
        setSize(new Dimension(200,50));
 }

 public void show() {
        super.show();
        super.toFront();
 }
}       // end of class
```

Listing A.4 Message.java

```
/*      Sun-Chong Wang
        TRIUMF
        4004 Wesbrook Mall
        Vancouver, V6T 2A3
        Canada
        e-mail: wangsc@triumf.ca

        Animate.java does the drawing on screen */

import java.awt.*;
import java.util.*;

class Animate extends Canvas implements Observer {
 double xmin,xmax,ymin,ymax;
 double topborder,sideborder;
 static int bottom, right;
 Dimension d;

 double[] spin;
 double[][][] site;
 int N;

 Graphics   os_g;
 Dimension  os_d;
 Image      os_i;
```

```
Spin parent;

Animate (Spin parent) {
        this.parent = parent;
        bottom = 500;
        right = 500;
}

public void update (Observable o, Object arg) { // from Observer
        Graphics g = getGraphics();
        re_display(g);
}

public void paint (Graphics g) {                    // from Canvas
        re_display (g);
}

public void re_display (Graphics g) {
        int i,j,k,l;
        int x0,y0,x1,y1;

        // this handles resizing of the window
        d = getSize();
        if ((os_g    == null)        ||
           (d.width  != os_d.width)  ||
           (d.height != os_d.height)) {
           os_d = d;
           os_i = createImage(d.width, d.height);
           os_g = os_i.getGraphics();
           bottom = d.height;
           right = d.width;
           SetScreenSize(right,bottom);
        }

        if (spin == null) N = 0;
        else N = (int) parent.sadlg.N;

        SetPlottingLimits();
        SetBorderSize(0.15,0.15);

        os_g.setColor(Color.white);
        os_g.fillRect(0,0,d.width,d.height);
        os_g.setColor(Color.black);

        // now plot
        os_g.setColor(Color.red);
        for (k=0; k<N; k++) {
            for (j=0; j<N; j++) {
                for (i=0; i<N; i++) {
                    l = i + j*N + k*N*N;
                    if (spin[l] > 0) {
                       os_g.setColor(Color.red);
                    } else {
                       os_g.setColor(Color.blue);
                    }
                    x0 = GetXCoordinate(site[l][0][0]);
                    y0 = GetYCoordinate(site[l][0][1]);
                    x1 = GetXCoordinate(site[l][1][0]);
                    y1 = GetYCoordinate(site[l][1][1]);
                    os_g.drawLine(x0,y0,x1,y1);
                }
            }
        }
        g.drawImage(os_i,0,0,this);
}       // end of re-display method
```

```
private int GetYCoordinate(double dValue) {
       int y = (int) ((1-topborder)*bottom-(1.0-2*topborder)*
                     bottom*(dValue-ymin)/(ymax-ymin));
       return y;
}

private int GetXCoordinate(double dValue) {
       int x = (int) (right*((1-2.*sideborder)*(dValue-xmin)/
                     (xmax - xmin))+right*sideborder);
       return x;
}

private void SetScreenSize(int x, int y) {
       right = x;
       bottom = y;
}

private void SetPlottingLimits() {
       xmin = (double) -N;
       xmax = (double) N;
       ymin = (double) -N;
       ymax = (double) N;
}

private void SetBorderSize(double fraction_of_x,
                          double fraction_of_y) {
       if ((fraction_of_x <= 0) || (fraction_of_y <= 0)) return;
       else {
            topborder = fraction_of_y;
            sideborder = fraction_of_x;
       }
 }
}      // end of Animate class
```

Listing A.5 Animate.java

Index

3-dimensional plot, 62
  coordinate transformation, 63
  subtended angle, 62
Acceptance-rejection method, 119–120, 184, 186
  2 dimensional distribution, 120
Action (in classical mechanics), 169–170
Akaike information criterion, 214
Animation, 63
Annealing, 59
Antithetic variable, 130
Antithetic variate method, 130
Argument passing, 242–243
Artificial neural network, 81
Bayes theorem, 195, 200, 215
Bayesian
  belief network, 210
  inference, 195–196, 210
  Kalman filter, 218
  learning, 210
  likelihood function, 196, 209
  maximum a posteriori, 196
  posterior probability, 196
  prior probability, 196
Bayesian analysis
  data, 210
  deconvolution, 200
  model selection, 195
Bayesian information criterion, 212, 214
Belief network, 210
Bivariate distribution, 121
Bivariate Gaussian, 122
Black-Scholes-Merton equation, 171
Boltzmann factor, 62
Boltzmann's constant, 60, 137
Boltzmann weight, 104
Brownian motion, 123, 172
Buffered I/O, 40
C, 241, 251
  array, 243, 251
  character, 242
  compiler, 242
  header, 248
  pointer, 243
  structure, 243
Cash flow, 123
Cellular automata, 147–149, 164–165
  hydrodynamics, 149
  lattice site, 148
    state, 148
  time, 148
  updating rules, 149
CERN, 241, 252
Chaotic behavior, 147
Chi-square, 181–182, 197–198, 208
  function, 77, 183, 202, 214
  minimization, 198, 201, 211
    Levenberg-Marquardt method, 209
    non-linear, 182
Chi-square fit, 181, 186, 188, 193, 200, 202, 221, 238
  2-dimensional, 193
  curvature matrix, 183
  estimation uncertainties, 183
  goodness of fit, 182
  gradient-expansion method, 183
  gradient method, 192
  Levenberg-Marquardt method, 193
  Marquardt method, 188
  number of degrees of freedom, 182
  steepest descent method, 192
Chromosome, 69, 101
Classical mechanics, 133–134
  acceleration, 133
  force, 133
  Newton's equations of motion, 134–135, 139
  Newton's second law, 133–134
Complexity, 147–148, 165
  self-organized criticality, 148

self-regularity, 148
self-similarity, 148
Computer experiment, 133
Conservation law, 122
Control variate method, 130
Convolution, 204
Coordinate transformation, 63
Correlation coefficient, 123
Correlation matrix, 121
Covariance matrix, 121
Critical exponent, 61
Critical temperature, 61–62
Cumulative probability function, 118
Data fitting, 77
figure of merit, 181
De Broglie thermal wavelength, 134
Deconvolution, 200–202, 208
Delta function, 199
Deoxyribonucleic Acid, 101
Derivative
numerically, 188
Dialog box
3-dimensional rendering, 73
financial options by path integral, 174
input format, 26
Kalman tracker, 235
Kohonen self-organizing map, 94
lattice gas automata, 152
molecular dynamics, 139
open file, 25
printing, 37
save file, 27
simulated annealing, 63, 69
stochastic-volatility jump-diffusion process, 125
TSP by genetic algorithm, 112
Distributed computing, 39, 41, 45, 54
Distribution, 117
2 dimensional, 120
bivariate, 121
bivariate Gaussian, 122
correlated multivariate Gaussian, 131
Gaussian, 78, 119, 121, 123
jump sizes, 123–124
mean of, 120, 197
multivariate, 121
normal, 119, 121
Poisson, 120, 123
standard deviation of, 197
uniform, 78, 119
univariate, 121
variance of, 121
DNA, 77, 101
Eigenvalue, 170
Ensemble average, 135
Entropy, 60, 196
Ergodic hypothesis, 135
Error, 120
of mean, 121
statistical, 120
systematic, 120
Error function, 69
Evolution, 101
Evolving artificial neural network, 116
Evolving neural network, 115
Ferromagnetism, 61
Feynman-Kac formula, 170
Feynman's path integral, 167, 180
Monte Carlo, 173–174
propagator, 168–170, 173
finite-time, 168–169
short-time, 168–169, 174
time evolution operator, 168
File input/output, 24
Financial options, 171
Black-Scholes-Merton equation, 171–172
Black-Scholes-Merton-Schrodinger equation, 172
Brownian motion, 172
Greeks, 174
log-normal distribution, 172
portfolio, 171
pricing, 171
propagator formalism, 172
risk-free interest rate, 171
strike price, 171
Wiener process, 172
Finite size correction, 138
Finite size effect, 138
Finite time correction, 138
Font, 36
Fortran, 241, 251
argument passing, 243
array, 243
common block, 243
compiler, 243
function, 242
library, 244
routine, 241–244, 250
Gaussian distribution, 119, 121, 123, 197
Gaussian integral, 172
Gauss-Jordan elimination, 10
Gene, 101
Genetic algorithm, 69, 101, 103, 112, 115, 214
co-evolution, 114
cooperation, 114
crossover, 102–103
crossover rate, 103
diversity, 103–104
mutation, 103
mutation rate, 103
objective function, 101, 104
selection, 104
tournament method, 104–105, 113
traveling salesman problem, 105, 115

Genetic programming, 114–116
crossover, 114
mutation, 114
non-terminal component, 114
terminal component, 114
tree structure, 114
Genome, 101, 148
Global positioning system, 214
Graph
arc, 211
directed graph, 211
node, 211
Graphical model, 211, 214, 238
learning, 211
Graphical user interface, 24
Grid computing, 39
Ground state, 170
Hamiltonian, 61, 168
Heisenberg uncertainty principle, 167
Helix, 218–219
High-performance computing, 39
H infinity filter, 211, 235, 239
gain matrix, 236
measurement equation, 235
measurement error, 235, 237–238
process disturbance, 235, 237–238
properties of, 236
Raccati difference equation, 236
system equation, 235
H infinity smoother, 237
Html, 253
HTTP web server, 44, 51
Impact parameter, 121
Importance sampling, 172, 174, 184
Internet, 39, 41
Inverse problem, 199
Inverse transform method, 118–119, 130
Ising model, 61, 63, 79, 103
3 dimensional, 63, 78
ground state, 62
Hamiltonian, 61
periodic boundary conditions, 62
regular cubic lattice, 62
spin, 61
Java
args[], 24, 44
argument pass by copy, 45
argument pass by reference, 9, 45
argument pass by value, 9, 45
array, 8, 10
1 dimensional, 69
2 dimensional, 45
3 dimensional, 69
args[], 11
index, 9
length, 10
bitwise operation, 152
buffered I/O, 40, 221
bytecode, 14
class, 3–4, 15
abstract, 28
ActionEvent, 24
Applet, 253
BasicStroke, 35
BorderLayout, 24
BufferedReader, 27
Canvas, 28
Color, 35, 90
Component, 28
FileInputStream, 26
Frame, 23
GeneralPath, 36
Graphics, 27
Graphics2D, 35
InputStreamReader, 26
Math, 14
Menu, 24
MenuBar, 17
Method, 51
Object, 13, 28, 45
Observable, 63
Panel, 23, 253
StreamTokenizer, 26
String, 45
Thread, 40, 55, 223
Throwable, 10
URL, 44
Vector, 39
class constructor, 9, 13, 23, 185
class loading, 42
dynamic, 42
class path, 52–53
color, 35
comment, 4
deep copy, 222
exceptions, 10, 13, 26, 48
extends, 10, 15, 23, 44
field, 8, 13, 23
file input/output, 24
final, 13, 35
for loop, 27
garbage collection, 10, 27
implements, 23–24
import, 4, 23
inheritance, 10, 15, 23
instantiation, 3, 8
interface, 23, 45, 48, 53
ActionListener, 23–24
Cloneable, 222
Observer, 63
Printable, 36–37
Remote, 44
Runnable, 40
Serializable, 48

jar, 253
java, 14, 52–53
javac, 14, 52–53, 242, 248
javah, 242, 248
Java native interface technology, 241
Matrix class, 63
message box, 27
method, 8–9
  actionPerformed(), 24–25
  add(), 24
  addActionListener(), 24
  clone(), 222
  close(), 27
  constructor, 9–10, 15
  drawString(), 36
  getActionCommand(), 24
  getDirectory(), 25
  getFile(), 25
  init(), 253
  main(), 3, 11, 13, 23–24, 44, 51, 253
  pack(), 23
  paint(), 28, 36–37
  print(), 37
  run(), 40, 186
  setFont(), 23
  setLayout(), 24
  setMenuBar(), 24
  setSize(), 23
  setStroke(), 36
  start(), 253
  System.out.println(), 10
multi-thread programming, 224
native, 242
native method, 241, 250
new, 8–9, 13, 15
object, 3–4
  abstract, 23
  font, 23
  string, 10, 24
package, 4, 53
  awt, 18
  util, 63
primitive data type, 8, 45
  boolean, 8
  byte, 8
  char, 8
  double, 8
  float, 8
  int, 8
  long, 8
  short, 8
public, 4, 9
recursion, 106
recursive method, 106
reflection, 42, 44, 51, 54
Remote Method Invocation, 41, 54
RMI, 41, 44
  client, 42
  rmic, 52
  rmiregistry, 52
  server, 42, 48
  stub, 44
security manager, 44, 51
serialization, 42, 48
statement, 4
static, 11
string, 242
super, 11
synchronized, 223
this, 23–24
thread, 40, 186, 192, 202, 221
throws, 10
try-catch, 10, 13, 26
Unicode, 242
variable field, 185
void, 11
while, 27
windowed programming, 17, 40
Java compiler, 14
Java Development Kit, 4
Java interpreter, 14
Java Native Interface, 241
Java virtual machine, 45, 250
Kalman filter, 211, 214, 238
  estimate error covariance matrix (a posteriori), 217
  estimate error covariance matrix (a priori), 216, 218
    propagation, 217
  gain matrix, 216–217
  measurement equation, 214, 221
  measurement error, 214, 216–217, 238
    covariance matrix, 217
  process noise, 214, 216–217, 221, 238
    covariance matrix, 217
  recursive procedure, 218, 238
  system equation, 214, 217
Kalman smoother, 217
  gain matrix, 218
Kohonen neural network, 89–90
  learning, 88–89
    unsupervised, 90
  self-organizing, 90
Kohonen self-organizing map, 88, 99
  clustering, 88, 92
  input layer, 88
  learning, 92–93
  learning rate, 90
  output layer, 88, 90
    2 dimensional, 99
  output neuron, 94
  weight, 92
Lagrangian (in classical mechanics), 170
Lagrangian multiplier, 197

Lattice gas automata, 149–150, 164–165
  collision rules, 150
  collision step, 151
  exclusion rule, 149
  face-centered hyper-cubic lattice, 150
  no-slip boundary condition, 152
  periodic boundary condition, 152
  transportation step, 151
  triangular lattice, 149–150
  updating rules, 149
Least squares method, 182
Lennard-Jones potential, 136, 139
Levenberg-Marquardt method, 183, 193–194, 202
Likelihood function, 196–197, 209
Logistic function, 81
Magnetic moment, 61
Magnetization, 61
Markov process, 215
Marquardt method, 188, 194
Matrix
  correlation, 121
  covariance, 121
  diagonal, 121
  inversion, 10, 218
  Java class, 4
  multiplication, 41
  triangular, 121, 123
Maximum a posteriori, 196
Maxwell-Boltzmann distribution, 138
Mean free path, 118
Metropolis algorithm, 60, 62, 69, 79
Metropolis-Hastings algorithm, 173, 180
Michel distribution, 185–187, 192, 196, 202
Michel parameter, 187–188, 196, 208–209
Michel spectrum, 202
Molecular dynamics, 133–135, 139, 144–145
  finite size effect, 144
  finite time effect, 144
  periodic boundary conditions, 138
  potential, 133
Momentum eigenstate, 169
Monte Carlo method, 117, 123, 183
Monte Carlo simulation, 117, 131, 134
  cash flow, 123
Moore's law, 39
Multi-dimensional integration, 173
Multi-processor system, 40–41
Multivariate distribution, 121
Natural selection, 101, 147
Navier-Stokes equation, 149–150
Neural network, 81, 86–89, 100
  activation function, 81–82
    arc tangent, 82
    hyperbolic tangent, 82, 86
    logistic, 82
    sigmoid, 82, 86
  architecture, 81, 88, 115
    feedforward, 84–85
    recurrent, 84, 87
  classifier, 89
  data representation, 86
  design, 86
  error function, 83, 85, 87–88
    squared errors, 87
  evolving, 115
  hidden layer, 81
    number of, 87
  input layer, 81
  Kohonen, 88–89
  learning, 86, 89
    supervised, 89
  neuron, 81,
    hidden, 87
    input, 81,
    output, 85, 87,
  node, 81
  output layer, 81, 88
  pattern recognition (structural), 84
  pattern recognition (temporal), 84
  recurrent, 84
  test (data) set, 83, 86
  testing, 86
  threshold, 81, 85
  time series prediction, 84
  training, 83, 86–87, 89
    (data) set, 83, 86
  transfer function, 81
  universal function approximator, 81
  validation, 86
    (data) set, 83, 86
  weight, 81, 85, 87–88
Newton's equations of motion, 134–135, 139
Newton's second law, 133
Normal distribution, 119, 121, 197
Normalization condition, 197
NP-hard, 113
Nyquist critical frequency, 181
Object oriented programming, 3, 15
Optimization, 101
Order parameter, 62
Parallel computing, 39, 41, 55, 105
  synchronization, 40
Path integral Monte Carlo, 173, 179
Periodic boundary conditions, 62, 138
Pixel, 200, 210
Pixel coordinates, 36
Pixon algorithm, 201, 208, 210
Planck's constant, 27, 168
Poisson counter, 125
Poisson distribution, 120, 123
Poisson statistics, 187
Portfolio, 88, 171
Position eigenstate, 168
Posterior probability, 196, 209–210

Power law, 148
Principle of maximum entropy, 196–197
Prior probability, 196, 200, 209
Probability distribution, 238
Propagator, 168
Quantum mechanics, 118, 134
  eigenstate (momentum), 169
  eigenstate (position), 168–169
  eigenvalue, 170
  ground state, 170
  Hamiltonian, 168
  Heisenberg uncertainty principle, 167
  probability, 168, 170
  probability amplitude, 172
  stationary action, 170
  wavefunction, 168, 170, 172
    completeness, 168
Quenching, 59
Raccati difference equation, 236
Random number generator, 117, 202
  cycle of, 117, 120
  Java class, 118
  pseudo, 118
  seed, 120, 131
  uniform, 117, 131
Recurrent neural network, 84, 87, 99, 115
Relaxation process, 174
Remote Method Invocation, 41
Response function, 199–200
Reynolds number, 150, 164
RGB, 90
RMI, 41
Rounding error, 149
Scale free, 148
Scale-free network, 148
Self-organized criticality, 148, 165
Self-regularity, 148
Self-similarity, 148
Semi-classical mechanics, 134
Shared Fortran library, 241
Sigmoid function, 81
Similarity, 92–94
Simulated annealing, 59, 62–63, 78, 88, 114–115, 201–202, 209, 214
  configuration, 59, 62, 69, 78
  cooling schedule, 62, 115
  critical temperature, 62
  global minimum, 59
  local minima, 59–60
  objective function, 59, 62, 77
  order parameter, 62
  transition probability, 60
Spin, 61
Spin glass, 69
Standard model, 196
Statistical error, 120–121, 192
Statistical mechanics, 135, 144–145
  ensemble, 135
  ensemble average, 135
  ergodic hypothesis, 135, 144
Stochastic process, 122
Stochastic volatility, 122
  mean reversion, 123
Stochastic-volatility jump-diffusion process, 117, 124–125
Symbiosis, 114
Systematic error, 120
Taylor series expansion, 137
Temperature, 60
Thermodynamics, 60, 79
Time series prediction, 84
Transitional probability density, 212
Traveling salesman problem, 59, 61, 69, 102–103, 105, 113
  NP-hard, 113
Triangular matrix, 121, 123
Truncation error, 137, 149
Uniform random number generator, 184
Univariate distribution, 121
Universality, 61
Variance reduction, 130–131
  antithetic variate method, 130
  control variate method, 130
Velocity Verlet algorithm, 137, 139, 144
Volatility, 122–123, 171
  stochastic, 122
Von Neumann method, 119, 184
Wavefunction, 134, 168
Web computing, 253
Wiener process, 123, 172
Wire chamber, 218–219, 221

MIX
Papier aus verantwortungsvollen Quellen
Paper from responsible sources
FSC® C105338

If you have any concerns about our products,
you can contact us on
ProductSafety@springernature.com

In case Publisher is established outside the EU,
the EU authorized representative is:
Springer Nature Customer Service Center GmbH
Europaplatz 3, 69115 Heidelberg, Germany

Printed by Libri Plureos GmbH
in Hamburg, Germany